Africa's demographic dynamics are rightly which will shape the Continent's path and that much needed focus on what that n farmers and entrepreneurs. No one could h to the fore the critical importance of African ownership and the centrality of those responsible for over three quarters of what Africa consumes: the local entrepreneur. Ndidi's life long-commitment and rigorous study of the issues pertinent is all evident in this timely publication.

Dr. Donald Kaberuka, 7th President, African Development Bank

Food Entrepreneurs in Africa is a unique book that sets forth the requirements to be a successful African change maker at each level of the vertical food chain from seed to the kitchen table. Ndidi Okonkwo Nwuneli emphasizes that change makers be more creative in responding to the health, economic, and environmental needs of both producers and consumers. To my knowledge, I know of no other book that has done this for Africa. This book will be useful to entrepreneurs, public policy makers, educators, and to all those who want to improve this much needed and too long neglected important part of the global food system.

Professor Ray Goldberg, Emeritus George M. Moffett Professor of Agriculture and Business, Harvard Business School

This book, through the voices of active entrepreneurs, distills the building blocks necessary to fortify courage for the brave few who will dare to rise to the challenge of feeding Africa, and the world. Through eight chapters, readers are carried along on a journey of discovery; parsing the challenges, motivations and value proposition for embarking on agribusiness entrepreneurship. This should be the first reference material for anyone, in the public or private sector, willing to contribute to Africa's prosperity in agribusiness. I am proud to recommend this book to governments, entrepreneurs and students of agribusiness. Mrs. Ndidi Okonkwo Nwuneli continues to demonstrate her leadership in fostering agriculture transformation in Africa.

Dr. Debisi Araba, Managing Director, African Green Revolution Forum (AGRF)

Food Entrepreneurs in Africa is a must-read for anyone who cares about building a more inclusive, fair and sustainable world, both within food and agriculture and beyond. Ndidi Okonkwo Nwuneli is an exceptional entrepreneur and story-teller. She offers a rare combination of hard won, on-the-ground practical experience, intellectual and analytical rigor and a deep-seated passion and personal commitment for driving transformational change. She knows first-hand how difficult it is to implement, let alone scale new financing and business models, new technologies and new mindsets – both at the firm-level and more systemically. She also understands the enormous positive potential when such change can be achieved. This book is both a practical guide for action and how to overcome obstacles as well as an inspiring vision and reflection on what is possible.

Jane Nelson,
Director of the Corporate Responsibility Initiative at Harvard Kennedy School, and a nonresident senior fellow at the Brookings Institution

Ndidi Nwuneli describes extensively in this book the need for scalable entrepreneurship in the agri-food industry in Africa. This is critical to make Africa self-supporting in its own food-production and more independent regarding food import and aid programs.

By developing knowledge and experience, Africa can become more resilient with respect to the adverse climate impact, natural disasters like locusts or infection diseases. This book is very welcomed and enhances the entrepreneurial spirit in Africa.

Feike Sijbesma, *Honorary Chairman Royal DSM, Africa Improved Foods, Global Climate Adaptation Centers*

FOOD ENTREPRENEURS IN AFRICA

Entrepreneurs are the lifeblood of the agriculture and food sector in Africa, which is projected to exceed a trillion dollars by 2030. This book is the first practical primer to equip and support entrepreneurs in Africa through the process of starting and growing successful and resilient agriculture and food businesses that will transform the continent. Through the use of case studies and practical guidance, the book reveals how entrepreneurs can leverage technology and innovation to leapfrog and adapt to climate change, ensuring that Africa can feed itself and even the world. The book will:

- Inspire aspiring entrepreneurs to start and grow resilient and successful businesses in the agriculture and food landscapes.
- Equip aspiring and emerging entrepreneurs with practical knowledge, skills, and tools to navigate the complex agriculture and food ecosystems and develop and grow high-impact and profitable businesses.
- Enable aspiring and emerging entrepreneurs to develop scalable business models, attract and retain talent, leverage innovation and technology, raise financing, build strong brands, shape their ecosystem, and infuse resilience into every aspect of their operations.

The book is for aspiring and emerging agribusiness entrepreneurs across Africa and agribusiness students globally. It will also inspire policymakers, researchers, development partners, and investors to create an enabling and supportive environment for African entrepreneurs to thrive.

Ndidi Okonkwo Nwuneli is a social innovator and entrepreneur with more than 25 years of experience. She is the cofounder of Sahel Consulting and AACE Foods, as well as the founder of LEAP Africa and Nourishing Africa. A graduate of the Wharton School and Harvard Business School, she serves on the boards of the African Philanthropy Forum, AGRA, Godrej Consumer Products, Nigerian Breweries, and the Rockefeller Foundation.

FOOD ENTREPRENEURS IN AFRICA

Scaling Resilient Agriculture Businesses

Ndidi Okonkwo Nwuneli

Routledge
Taylor & Francis Group
LONDON AND NEW YORK

First published 2021
by Routledge
2 Park Square, Milton Park, Abingdon, Oxon OX14 4RN

and by Routledge
52 Vanderbilt Avenue, New York, NY 10017

Routledge is an imprint of the Taylor & Francis Group, an informa business

© 2021 Ndidi Okonkwo Nwuneli

The right of Ndidi Okonkwo Nwuneli to be identified as author of this work has been asserted by her in accordance with sections 77 and 78 of the Copyright, Designs and Patents Act 1988.

All rights reserved. No part of this book may be reprinted or reproduced or utilised in any form or by any electronic, mechanical, or other means, now known or hereafter invented, including photocopying and recording, or in any information storage or retrieval system, without permission in writing from the publishers.

Trademark notice: Product or corporate names may be trademarks or registered trademarks, and are used only for identification and explanation without intent to infringe.

British Library Cataloguing-in-Publication Data
A catalogue record for this book is available from the British Library

Library of Congress Cataloging-in-Publication Data
Names: Nwuneli, Ndidi Okonkwo, author.
Title: Food entrepreneurs in Africa: scaling resilient agriculture businesses / Ndidi Okonkwo Nwuneli.
Description: First Edition. | New York: Routledge, 2021. | Includes bibliographical references and index.
Identifiers: LCCN 2020042891 (print) | LCCN 2020042892 (ebook) | ISBN 9780367631079 (hardback) | ISBN 9780367631123 (paperback) | ISBN 9781003112167 (ebook)
Subjects: LCSH: Food industry and trade–Africa. | Agricultural industries–Africa. | Entrepreneurship–Africa.
Classification: LCC HD9017.A2 N98 2021 (print) | LCC HD9017.A2 (ebook) | DDC 338.1096–dc23
LC record available at https://lccn.loc.gov/2020042891
LC ebook record available at https://lccn.loc.gov/2020042892

ISBN: 978-0-367-63107-9 (hbk)
ISBN: 978-0-367-63112-3 (pbk)
ISBN: 978-1-003-11216-7 (ebk)

Typeset in Joanna
by Deanta Global Publishing Services, Chennai, India

To my wonderful siblings – Adaora, Una, Nwando, and Aneto – your love, support, friendship, and guidance have been priceless. You all are God's greatest gift to me.

To the heroes of the COVID-19 pandemic – our frontline workers in the health and food ecosystems, especially our teams and partners at AACE Foods, Sahel Consulting, Sahel Capital, and Nourishing Africa. Thank you for your sacrifices to sustain lives on the African continent!

CONTENTS

List of Illustrations ... xi
About the Author ... xiv
Foreword ... xvi
Acknowledgments ... xviii
Introduction ... xxii

1 Overview of the Food and Agriculture Landscape in Africa ... 1

2 Developing a Compelling and Sustainable Business Model ... 12

3 Talent For Scaling ... 29

4 Leveraging Technology and Innovation to Scale Your Business ... 57

5 Building Your Brand and Amplifying Your Impact ... 83

6 Financing Your Growth ... 110

7	Shaping Your Ecosystem	134
8	Building Resilience – Adapting to Climate Change, Mitigating Risks, and Shocks	160
	Conclusion	174
	Index	177

ILLUSTRATIONS

Figures

1.1	Sahel Consulting's overview of the pain-points in the African agriculture landscape	4
2.1	Sahel Consulting's maize value chain in Nigeria	16
2.2	Prerequisites for business models that scale in the African agriculture and food landscape	24
3.1	SoFresh's recruitment process	42
3.2	SoFresh 2020 organizational chart	43
3.3	Performance measurement indicators utilized by Good Nature Agro, Zambia	52
4.1	Spectrum of digital interventions on the African continent	58
5.1	The six Ps	92
5.2	AIDAR model	104
7.1	Domains of the entrepreneurs' ecosystem	135
7.2	Growing better: Ten critical transitions to transform food and land use	150

Tables

4.1	Examples of Players in the Agtech/Digital Ag Ecosystem	61
5.1	Examples of African Agribusinesses and Their Social Media Footprint	102

5.2	Measuring the Effectiveness of Your Marketing Strategy	107
6.1	Financing Options by the Stage of Your Business Life Cycle	111
6.2	Examples of Agribusinesses that have Benefited from Competitions, Fellowships, Prizes and Awards	116
6.3	Key Issues Assessed by Providers of Capital	121
6.4	One Acre Fund's Fundraising Journey	129

Boxes

1.1	Cocoa in Ghana	6
2.1	Consider AACE Foods team's responses to similar questions	21
2.2	Examples of mission and vision statements and values	22
3.1	Typical roles in agriculture and food businesses	38
3.2	IITA GoSeed – Outsourcing	46
4.1	Case study – Early Generation Seed (EGS) Systems for the cassava value chain	63
4.2	Biofortification – HarvestPlus' pioneering work in seed systems	64
4.3	OCP's pioneering efforts in the African fertilizer landscape	66
4.4	Logistics in Africa – Kobo360	73
5.1	Selecting a brand name for AACE Foods	85
5.2	Maneli Pets' brand strategy	88
5.3	Case study – Brookside Dairy Ltd, Kenya	99
5.4	Two distinct approaches to brand building	105
6.1	The Fund for Agricultural Finance in Nigeria (FAFIN)	119
6.2	Reasons VC/PE funds decline applications from agrifood SMEs	123
6.3	Raising funds – the ReelFruit experience	126
7.1	Nourishingafrica.com – A digital home for entrepreneurs in the African food and agriculture landscape	142
7.2	Case study – AACE Foods' new factory	144
7.3	Case study: Africa Improved Foods Rwanda	146
8.1	Snapshot of company resilience in April/May 2020 during COVID-19 from the NourishingAfrica.com hub	162

Worksheets

2.1	Matrix for Finding Your Sweet Spot	17
2.2	Determining Your Business Model	21
2.3	Developing or Refining Your Mission, Vision, and Values	23
2.4	Assessing Whether Your Business Model is Demand-Driven	25
2.5	Measuring Impact	27
3.1	Assessing Your Leadership Skills	32
3.2	How Effective Is Your Board?	34
3.3	How Effectively Are You Attracting Star Employees to Your Company?	41
3.4	Mapping Your Team	53
4.1	Assessing Your Company's Use of Agtech, Digital Tools, and Innovation	60
5.1	Clarifying Your Branding Positioning	93
6.1	Investment Readiness	125
7.1	Map Your Ecosystem – Insert Specific Names of Individuals or Individuals Who Fit within Each Category	137
7.2	Assessing Your Regulatory Environment	138
7.3	Assessing Your Competitive Landscape	140
7.4	Assessing the Geography and Infrastructure Structures in Your Ecosystem	141
7.5	Assessing the Associations in Your Ecosystem	143
8.1	How Resilient Is Your Business?	164
8.2	Assessing Your Financial Resilience	169

ABOUT THE AUTHOR

Mrs. Ndidi Okonkwo Nwuneli is an expert on African agriculture and nutrition, philanthropy, and social innovation. She has more than 25 years of international development experience and is a recognized serial entrepreneur, author, public speaker, and consultant.

Through Ndidi's work as the cofounder and the managing partner of Sahel Consulting Agriculture & Nutrition, she has partnered with development partners and a range of private- and public-sector organizations to shape policies and implement ecosystem solutions in the African agriculture and food landscapes. As the cofounder of AACE Foods, which produces a range of packaged spices, seasonings, and cereals for local and international markets, Ndidi has propelled the growth of a catalytic business. As the founder and chair of Nourisingafrica.com, a digital home for food and agriculture entrepreneurs operating on the African continent, she is accelerating the growth of small- and medium-size enterprises in the sector. Ndidi is also the founder of LEAP Africa, which works in six African countries, inspiring, empowering, and equipping a new generation of African leaders.

Ndidi serves on the boards of the Rockefeller Foundation, AGRA, Nigerian Breweries Plc. (Heineken), Godrej Consumer Products Ltd. India, Fairfax Africa Holdings Canada, Businessday Newspapers, and the African Philanthropy Forum.

She started her career as a management consultant with McKinsey & Company, working in Chicago, New York, and Johannesburg. She returned to Nigeria in 2000 to serve as the pioneer executive director of FATE Foundation, supporting young entrepreneurs to start and scale their businesses. In 2002, she established LEAP Africa to inspire, empower, and equip a new cadre of principled, disciplined, and dynamic young leaders in Africa.

Ndidi was recognized as a Schwab Social Innovator and Young Global Leader by the World Economic Forum and received a National Honor from the Nigerian Government. She is a TED Global speaker, and her work has been featured on CNN, BBC, and a range of international and local media outlets.

Ndidi holds an MBA from the Harvard Business School and an undergraduate degree with honors from the Wharton School of the University of Pennsylvania. She was a senior fellow at the Mossavar-Rahmani Center for Business & Government at the Harvard Kennedy School and an Aspen Institute New Voices Fellow.

Ndidi is the author of *Social Innovation in Africa: A Practical Guide for Scaling Impact*, published by Routledge in 2016

FOREWORD

From the first page of this brilliant book we are in author Ndidi Nwuneli's thrall. The author quickly reveals two realities that have shaped this book and her own professional life. First, Africans must own and drive the transformation of their food systems and, second, entrepreneurs have to be at the heart of the transformation. Here the author walks the talk.

The book is wonderfully written and researched. Data and evidence are carefully interspersed with inspiring narrative stories about the realities facing real entrepreneurs, the many opportunities that exist for them in every sector and in every value chain, and what they need to do to work out how to create and seize those opportunities and ride out the inevitable shocks.

In fact, the book is a veritable "how to" for entrepreneurs in the food sector – in any country – with guidance, worksheets, and examples. The chapter on leadership skills and self-assessment is unusual for a book of this kind, with the author reminding us that potential investors are investing in the entrepreneur as much as the business model. The proactive search for talent is another critical issue that is rarely written about, but Nwuneli brings the issue to life and illustrates how critical it is, underpinned by the creation of a high performance conducive organizational culture. Other topics include brand building, the use of technology, and, one of my favorite chapters (because of its clarity), financing growth.

Not content with making one's own business a success, Nwuneli also has a chapter on shaping your ecosystem, which highlights the broader agency of entrepreneurs working together in partnerships and networks to shape the business policy environment they face, which in turn generates more growth opportunities.

The book is a rare treat: It is grounded in the reality of a successful businessperson, a reality that is contextualized by a superb reach into the academic literature, enlivened by dozens of case studies and reflective in that it holds up a mirror to the reader and to potential and current entrepreneurs to help them self-analyze and strengthen their business.

Simply put, it is probably the best book I have ever read on this topic. I will make sure it is widely available to all of my staff at the Global Alliance for Improved Nutrition. I will urge my colleagues and partners to read it, learn from it, share it, critique it, and, critically, act on it. I urge you to do the same.

Dr. Lawrence Haddad
Executive Director, Global Alliance for Improved Nutrition (GAIN)
2018 World Food Prize Winner

ACKNOWLEDGMENTS

When God laid it on my heart to write this book, I was genuinely perplexed about how I would find the time given the many responsibilities that I was juggling as a serial social entrepreneur, wife, mother, daughter, sister, public speaker, and board member. However, I felt a sense of urgency to commence the process, with no clue that the COVID-19 pandemic would propel the world into another food crisis and further reinforce the critical need for this book.

I remain incredibly grateful to Scott Leland for welcoming me back to the Mossavar-Rahmani Center for Business & Government at the Harvard Kennedy School and for his passion, dedication, kindness, and support during this past year as a research fellow. I also thank the terrific team at the Center – Professor John Haigh, Professor Richard Zeckhauser, Susan Gill, and Victoria Groves-Cardillo. I recognize the wonderful guidance and support provided by my faculty advisors – Professor William Clark and Professor Zoe Marks – and for the mentorship provided by Jane Nelson. I acknowledge the pioneering work of the late Professor Callestous Juma, who urged me to write yet another book.

I am grateful to Dr. Lawrence Haddad for the excellent foreword and the fantastic global leaders who provided reviews for this book – Dr. Adebisi Araba, Professor Ray Goldberg, Dr. Donald Kaberuka, Jane Nelson, and Feike Sijbesma.

I would like to recognize the publisher, Routledge, especially the senior editor, Rebecca Marsh and the editorial assistant, Sophie Peoples.

I am grateful to the amazing faculty at the Harvard Business School for their support – Professor John Macomber, Professor Jose Alvarez, Professor Debora Spar, Professor Linda Hill, Professor Forest Reinhardt, and Pippa Tubman Armerding.

I am extremely grateful to my primary research associate, Ifeoma Umunna, for her passion and dedication, and the many dynamic young people who provided research support, including Rahmat Eyinfunjowo, Devika Balachandran, Iyala Alai, Steph Barker, Uyaietieno Okonnah, Bomi Fagbemi, Mayowa Omogbenigun, Ijeoma Nwuke, Kabir Singh, and Udenna Nwuneli. Thank you to Nanre Nafzinger for her editing support.

I am grateful for the incredible support provided by my colleagues at Sahel Consulting, Sahel Capital, Nourishing Africa, and AACE Foods. I especially recognize the insights and case studies provided by Mezuo Nwuneli, Temi Jebutu, Temi Adegoroye, Fisayo Kayode, Funke Okuwobi, Oladele Shekete, and Bolaji Etti.

I am incredibly grateful to the brilliant Dr. Akinwumi Adesina, Dr. Rajiv Shah, Dr. Agnes Kalibata, Dr. Audu Grema, Bonnie McClafferty, Carla Denizard, C.D. Glin, Dr. Gbenga Ogunmoyela, Jeff Dykstra, Professor Jessica Fanzo, Julie Howard, Kimberly Flowers, Kristen MacNaughtan, Larry Umunna, Lindiwe Sibanda, Mark Nelson, Myriam Sidibe, Roy Steiner, Dr. Shachi Gurumayum, Dr. Lauren Good, Lawrence Kent and Dr. Victor Ajieroh and for their thought leadership in the food, nutrition, and agriculture landscapes and for serving as excellent sources of inspiration and encouragement over the years.

I recognize Jeremy Oppenheim, Marshall Burke, Meron Demisse, and Neal Keny-Guyer, four brilliant and passionate individuals who spent two days with me in the beautiful Rockefeller – Bellagio Center, sharing insights and challenging my perspectives, and whose prodding led to the birth of Nourishingafrica.com.

I also thank the amazing entrepreneurs, investors, policymakers, funders, and development partners who graciously provided insights for the book. They include: Adan Mohammed of Ecodudu; Affiong Williams of ReelFruit; Alima Bawah of CowTribe; Amina Rashad of Get Your Glow; Angel Adelaja-Kuye of Fresh Direct; Akosua Ampofo Siever and William Foote of Root Capital; Andrew Gartside of DFID, Aubrey Hruby of Africa

Expert Network; Ayodeji Balogun of AFEX; Baafour Otu-Boateng of Investisseurs & Partenaires (I&P); Bethel Buzuayehu, Batuael Bizenu, and Egziael Buzuayehu of East Africa Holding; Bosun Tijani of CCHUB; Brian Frimpong of Zebu Investments, Chef Fatmata Binta and Chirene Jelbert of C-Fruit; Claire Van Der Kleij of Pep Zambia; Daniel Asare-Kyei of Esoko; Daniela Amosun and John Chitsa of TechnoServe South Africa; Dr. Tony Bello and Edward Mabaya of AfDB; Eleni Gabre-Madhin of Blue Moon; Ekenem Isichie of ACIOE; Evans Danso of Flosell LLC; Fred Zamble of Seekewa; Frederick Schreurs of IITA GoSeed Cassava; Gaurav Vijayvargiya of Seba Foods; Gillian Hutchings of National Business Initiative (NBI); Hadija Jabiri of GBRI Business Solutions LTD; Hillary Miller-Wise of Tulaa and Ike Ilegbune of iGate Consulting; Irene Ochem of AWIEF; Jeff Dykstra of Partners in Food Solution; Jehiel Oliver of Hello Tractor; Jolene Dawson of Accenture; Joseph Nkandu of NUCAFE; Joséphine Katumba of Biakudia Urban Farming Solutions; Joshua Adedeji and Tunji Ajani of ANDE West Africa; Jurgen Paulis and the entire team at Africa Improved Foods; Keith Bolshaw of Pacific Ring West Africa; Kenton Dashiell and Hemant Nitturkar of IITA; Khady Diop of WIN Industries; Kola Masha of Babban Gona; Lesego Serolong Holzapfel of Bokamoso Impact Investments; Lillian Secelela Madeje, Ekihya, and Micheline Ntiru of Stanford Seed Global Business School; Milton Lore of KPMG; Mira Mehta of Tomato Jos; Monica Musonda of Java Foods; Naana Winful Fynn and team of Norfund; Natalia Pshenichnaya and Rita Oulai of GSMA; Nhlanhla Dlamini of Nandi Pets; Nkiru Okpereke of Enviro-Gro; Nneka Eze of Dalberg; Nola Mashaba of Nola Communications; Obi Ozor of Kobo360; Olagoke and Abimbola Balogun of SoFresh Nigeria; Peris Bosire of FarmDrive; Peter Njonjo of Twiga; Ronke Aderinoye of AgriHub; Sara Boettiger of McKinsey; Seyi Abolaji of Wilsons Juice Company; Sipamandla Manqele of Local Village Foods; Stiaan Wandrag of Tiger Brands; Sunday Silungwe and Kellan Hays of Good Nature Agro; Sydney Zharare of DevPact; Temi Jebutu of AACE Foods; Theo Chiyoka of Traxi; Tosan Jemide of Top Crust Bakery; and William Asiko of The Rockefeller Foundation.

I thank my dear friends who provided emotional and spiritual support during my many trips to Cambridge – Cassandra Johnson, Isioma Chukwu, Pastor John and Ngozi Enelamah, Nanre Nafzinger, Professor Jacob Olupona, Raynya Simmons, Tonica Riley, and Uche Pedro.

Thank you to my dearest sisters and accountability partners, Laura Nneka Mobisson-Etuk, Eme Essien Lore, Toyosi Kolawole, Oge Modie, Pam Egbo, Nadu Denloye and my communities of support – the amazing Power Girls, the Praying Ladies, the HN sisters, Couples Fellowship, C3 – Covenant Nation, the HBS Praying Sisters, the APF, and the LEAP Africa families.

I appreciate my wonderful parents, Professors Paul and Rina Okonkwo, siblings, their spouses, and children for their unconditional love and support and for being my greatest champions.

I thank my husband, Mezuo, who has been my colaborer in the agriculture and food sector, and my children, Udenna and Amara, who have challenged me to live my best life.

Finally, I thank God – the source of my strength and the reason I live – for favor, grace, wisdom, and the conviction to contribute toward ensuring food security on the African continent.

INTRODUCTION

My childhood at the University of Nigeria Enugu Campus in Southeastern Nigeria, first nurtured my interest in agriculture. Our home had a mini orchard with mango, guava, avocado, and papaya trees, as well as roses, gardenias, and cherry hedges and a vegetable garden with cucumbers and green beans. We had enough to eat and share with our neighbors in a very close-knit community.

At 16, when I moved to the United States to complete high school and start college, I was shocked to learn that Africa's global face was a starving disheveled child with yellow eyes, pale skin, and an extended stomach. I did not recognize this child, and yet, this picture of starvation followed me around the world – at dinner parties, conferences, on television, or even in train stations. I recoiled against those stereotypes, and they propelled me to devote my life to transforming the global narrative about Africa.

Through my work as a social entrepreneur engaged in youth and leadership development in the late 1990s and early 2000s with the Ford Foundation, FATE Foundation, and LEAP Africa, I interfaced with many entrepreneurs in the African agriculture and food sectors. They shared their challenges with developing demand-driven and sustainable business models, finding and retaining talent, raising appropriate financing, building strong brands, and shaping their ecosystems. Many of these entrepreneurs struggled for survival in exceedingly difficult and often hostile environments. While I tried to provide training, support, and encouragement to them, I was ill-equipped to fully understand the magnitude of the hurdles they faced, until I started my agribusinesses.

Over the past 12 years, as I have immersed myself in this sector through AACE Foods, Sahel Consulting, and Nourishingafrica.com, I have recognized and embraced two realities:

First, no external force can fully address Africa's agriculture and food challenges. Africans must drive and own the transformation required in the sector. This means that even though Africa is naturally endowed for agricultural excellence, the fact that our region remains a net importer of food is unacceptable. We must urgently address the serious structural, infrastructural, equity, and skills gaps limiting us. We also must proactively adapt to the climate crises, which is further compounding our struggles, causing flooding, droughts, and elevated temperatures, and dramatically affecting agricultural ecosystems.

Second, entrepreneurs across the different value chains in Africa must be at the forefront of this transformation; they are the lifeblood of this sector. According to the Africa Agriculture Status Report 2019[1], published by the Alliance for a Green Revolution in Africa (AGRA), the African private sector handles 80% of the food consumed on the continent. Referred by AGRA as the "hidden middle," these companies are the drivers of growth and innovation. They provide inputs, logistics, processing, and distribution of critical food and cash crops, livestock, and all other support services.

These African entrepreneurs require support, knowledge, financing, and tools to scale their businesses and leverage the available technology and innovation to leapfrog. They must effectively position and prepare themselves to maximize the opportunities in this sector, projected to exceed $1 trillion[2] in market value by 2030.

Over the past year, as part of my research fellowship at the Mossavar-Rahmani Center for Business & Government at the Harvard Kennedy School, I interviewed more than 80 entrepreneurs, investors, policymakers, funders, and development partners to understand the critical steps required to build resilient agriculture and food businesses on the African continent. I have captured their unique insights, coupled with my experiences as a social entrepreneur in the eight chapters, which lay out practical strategies for aspiring entrepreneurs to start and grow successful businesses. The case studies and tools will also enable emerging entrepreneurs to navigate the complex agriculture and food ecosystems, develop scalable business models, attract and retain talent, leverage technology and innovation, raise financing, build strong brands, shape their ecosystem, and grow high-impact, resilient, and profitable businesses

The COVID-19 pandemic has reinforced the link between our health and food systems and the critical role that nutritious diets play in boosting our immune systems and ensuring human survival. This crisis has also exposed the fragility of our food ecosystem, with disruptions affecting national, regional, and international trade, limiting farmers from accessing inputs and markets, and leading to food shortages and price hikes. The rising levels of hunger and malnutrition have further strengthened our resolve to reimagine, retool, and rebuild a resilient, just, and sustainable food and agriculture landscape that can withstand future shocks.

I hope that this book will inspire, equip, and support African entrepreneurs to build successful agribusinesses and transform the landscape. It will also inspire policymakers, development partners, and researchers to create an enabling environment for agribusinesses to thrive, ensuring a new image of Africa that replaces the sad one I was confronted with in my youth. A happy child with big, beautiful brown eyes, healthy skin and hair, and a wide smile will show how well Africa can nourish itself and the world.

Notes

1. Alliance for a Green Revolution in Africa.(2019) *The hidden middle: A quiet revolution in the private sector driving agricultural transformation* [Online]. Available at: https://agra.org/wp-content/uploads/2019/09/AASR2019-The-Hidden-Middleweb.pdf (Accessed: 25 August 2020).
2. African Development Bank Group. (2018). *Africa agribusiness, a US$1 trillion business by 2030.* Available at: https://www.afdb.org/en/news-and-events/africa-agribusiness-a-us-1-trillion-business-by-2030-18678 (Accessed: 25 August 2020).

1

OVERVIEW OF THE FOOD AND AGRICULTURE LANDSCAPE IN AFRICA

The global food industry is large and complex. With an annual 5% growth rate, it is expected to rise to US$20 trillion by 2030[1], propelled by consumer demand for healthy food, convenience, and sustainability.

Africa, as a region, is gradually becoming an essential stakeholder in this landscape. With 1.3 billion people in 2020, Africa's population is predicted to double to 2.5 billion by 2050[2]. This rapid population growth presents many challenges and opportunities for the continent's food systems and agricultural sectors. The anticipated increase in demand for food introduces greater entrepreneurial and investment opportunities in key subsectors.

However, there are a few critical realities and promising trends in the food and agriculture ecosystem, which must be understood and addressed to maximize the opportunities fully in this vital and vibrant sector.

Critical realities

There are at least eight "critical realities" that currently characterize the food and agricultural landscape in Africa.

1. **A fragmented ecosystem still dominated by smallholder farmers:** Cash and food crop production is dominated by smallholder farmers on one end of the spectrum and large, multinational trading and processing companies on the other end. Aggregators and middlemen extract value between these two extremes. Unproductive, fragmented, and unstructured value chains, with prohibitive production costs; limited use of improved seeds, fertilizer, and irrigation; and poor storage and processing make most value chains uncompetitive relative to their international counterparts.
2. **Climate change:** This is a critical reality facing the food and agriculture ecosystem in Africa. Of the ten countries considered most threatened by climate change globally, nine[3] are in Africa. The recent floods, droughts, and locust infestations that have plagued most of the continent over the past few years reinforce the reality that climate change can completely reverse the advancements in the agriculture and food ecosystem unless urgent action is taken.
3. **High rates of postharvest losses and limited local processing:** There are high levels of food waste estimated at between 20% and 60%[4], depending on the value chain. This is linked to the underdeveloped local food manufacturing and processing landscapes, high cost of transportation, fragmented distribution and retail channels, with the vast majority of food is still sold in the informal open-air markets and there are limited cold chain networks. This has led to a dependency on imported processed foods. These dynamics result in more than 50% of household income being spent on food in most African countries. The one exception is South Africa, which spends 19.1%, relative to Nigeria's 56.6%. This is much higher than the United States at 6.4%, the United Kingdom at 8.2%, India at 29%, and Brazil at 15.6%[5].
4. **High rates of malnutrition:** According to UNICEF's *State of the World's Children 2019* report, undernutrition, micronutrient deficiencies, and overweight (obesity, in its most severe form) are the three strands of the triple burden of malnutrition. At least one in three African children is stunted, which affects not only their ability to grow to their full potential but also their brain development. Sadly, in the crucial first 1,000 days, the majority of children are not getting the nutrition they need. As a result, even though an increasing number of children

and young people are surviving, far too few are thriving because of malnutrition[6].

Overweight and obesity, historically believed to only affect the wealthy, are now increasingly a condition of the poor, given the greater availability of fatty and sugary foods around the world[7]. This is also leading to higher rates[8] of noncommunicable diseases, such as diabetes and high blood pressure, among the adult population in Africa.

Despite government efforts to address micronutrient deficiencies through mandatory fortification initiatives, compliance has been relatively low. In addition, poor food choices and the limited availability, affordability, accessibility, and acceptability of nutritious food, especially for low-income and vulnerable populations, have further exacerbated the high rates of malnutrition.

5. **Significant infrastructure, talent, and financing gaps:** Africa remains energy-poor, with more than 600 million households and many businesses without access to electricity. This directly increases the cost of doing business and the growth of the agriculture and food sectors. While there are many innovations and off-grid solutions that leverage renewable energy sources, the pace of adoption is slow. In addition, because of poor transportation infrastructure, most rural communities are not accessible by trucks and cars, and there are limited linkages between farms and markets. For example, in Nigeria, one of Africa's biggest economies, only 25%[9] of the rural villages are currently accessible via tarred roads.

There is also a shortage of talent for driving growth and innovation in the sector. Sahel Consulting's Quarterly[10] research reveals that most agriculture programs in African universities struggle to attract students. The curriculum is outdated, and there are limited links between theory and practice. For governments that manage extension services programs, the ratios of extension workers to farmers is as low as 1:8,000[11], and these workers often do not have the latest training or technology, which, in turn, limits knowledge transfer and the effectiveness of their interventions.

Despite recent positive initiatives to promote financing in the African agriculture and food landscapes, entrepreneurs still cite access to affordable and patient growth capital as the most significant constraint. This is especially evident in the $50,000–$250,000 funding window, which investors claim requires the same amount of due diligence and monitoring time as much larger amounts, while generating smaller returns with more higher risk exposure.

6. **Poor enabling and regulatory environment and weak enforcement of standards:** Many countries still struggle with limited data-driven policymaking, delays in implementation of policies, abrupt reversals or contradictory policies, inadequate management capacity for the implementation of standards, and ineffective trade regulations. This limits the effectiveness and efficiency of entrepreneurs and frustrates investors, deterring the influx of growth capital. It also fosters unethical behavior and rising rates of food fraud[12], a term used to describe substandard food sold in many markets that has been altered, misrepresented, mislabeled, and substituted.

7. **Gender inequity:** According to the UN Women[13] report *The Gender Gap in Agricultural Productivity in Sub-Saharan Africa: Causes, Costs and Solutions*, significant gender gaps persist that directly impact productivity, livelihoods, and poverty levels in Africa. In fact, the report revealed gender productivity gaps between 11% in Ethiopia and 28% in Malawi, which are linked to a range of constraints, including limited access to fertile land, inputs such as improved seeds and fertilizer, and mechanization support. Other gender gaps exist in agriculture research and development, land ownership, agricultural extension services, access

Figure 1.1 Sahel Consulting's overview of the pain-points in the African agriculture landscape. Source: Sahel Consulting Agriculture and Nutrition Limited.

to financing, aggregation, distribution, logistics, and the production and processing of high-value crops.

Closing the gender gap in Africa will have a transformational impact not only on the most vulnerable households but also on the entire ecosystem within countries and across regions. For example, according to the estimates shared in the UN Women report, addressing the productivity gap alone in Rwanda will increase crop production by 19% and GDP by more than $419 million and lift more than 238,000 people out of poverty over ten years.

8. **Regional and global trade dynamics:** Africa remains a net importer of food, spending between $45 billion and $50 billion in 2019 on imports, exporting between $35 billion and $40 billion with only $8 billion in interregional trade[14]. The continent is also a large exporter of cocoa, coffee, tea, sesame, and cashew to the world[15]. Many countries contend with straddling a focus on food crops versus cash crops, given the risks of exposure to highly volatile commodity markets and competing with other economies that offer significant subsidies to their farmers.

   The pace of globalization has increased, and national economies have become ever more interconnected; however, the patterns of trade in Africa still echo the patterns that characterized African economies under colonial rule. In many instances, food producers still export "raw materials" such as unprocessed coffee and cocoa to be processed and packaged in other countries. In some cases, consumers even "buy back" products such as chocolate bars and instant coffee that were processed in different countries and then imported as manufactured goods. These processed goods often generate much higher prices than the raw material that went into making them. This usually leaves producers and consumers at either end of the supply chain unable to capture the added value that comes with processing and manufacturing.

   There is also tremendous inequity in global trade and double standards with Africa. This includes different standards for food imports and exports, depending on the country of origin, with the greater burden placed on African entrepreneurs to prove themselves repeatedly, to enter Europe and the United States, and lower standards for imports into the continent.

Box 1.1: COCOA IN GHANA

For decades Ghana has been one of the largest producers of cocoa in the world. Today, Ghana and the Ivory Coast produce almost two-thirds of the global supply of cocoa – the main ingredient in what is now a $100 billion industry. Unfortunately, farmers and local producers rarely reap their share of these profits. Ghana supplies about one-fifth of the world's cocoa beans. It earns about $2 billion a year – less than 2% of the value of chocolate that is ultimately manufactured, branded, and sold as a finished product. The average cocoa farmer in Ghana can expect to earn between $0.40 and $0.45 per day, or between $983 and $2,627 per year. In contrast, the Mars family, which owns Mars Inc. – the makers of M&M's and Snickers, Milky Way, and Twix bars – has a net worth of $90 billion, making them the third richest family in America.

Over the past few years, the chocolate industry has come under increased pressure to develop more just and equitable supply chains. In response, the fair-trade industry has grown into a $9 billion industry. While the increased scrutiny of supply chains is a promising trend, the shift toward "fair trade" will not be enough to break out of these historical import–export cycles. Farmers and producers on the continent will not be able to capture their fair share of the market value until a greater share of finished products are manufactured, branded, and sold closer to the source of their raw materials.

A few entrepreneurs have tried to make chocolate in Ghana. Many have faced insurmountable difficulties, such as the absence of a substantial dairy industry, high electricity prices, and lack of refrigeration facilities to keep the chocolate cool in Ghana's tropical climate. In the past, these factors have made it challenging to manufacture at origin.

Another challenge facing the industry is that few children want to take up their parents' farms when they come of age. The average cocoa farmer in Ghana is 60 years old. Urbanization has been increasing dramatically over the past few decades.

This case study of cocoa production in Ghana illustrates how many critical realities and promising trends can overlap and intersect around one food type or one entrepreneurial venture. Individuals interested in working in the cocoa industry may have to simultaneously evaluate and manage challenges and opportunities that include increasing population growth, international trade, climate change and deforestation, historical inequities in the supply chain, and an aging farming population.

Promising trends

Many promising trends are noteworthy, six of which are highlighted below.

1. **Momentum of digital innovations**: According to CTA's *Digitalisation of African Agriculture Report, 2018-2019*[16], there are at least 390 distinct, active "digitalization for agriculture" solutions across the continent. They are expected to be game-changers in the food and agriculture landscapes. They can propel precision agriculture, enhance yields, reduce the cost of food production, ensure effective planning and communication, reduce postharvest losses, provide linkages to insurance and financial markets, and create cohesive value chains and efficient markets. While some of these interventions are reaching hundreds of farmers, others have low adoption rates because of their relatively high cost of usage and deployment challenges. Many of these innovations can predict droughts and help farmers and stakeholders across critical value chains institute climate adaptation strategies. Kenya already has more than 80% of its farmers registered on these platforms. It is also home to 20 of the most comprehensive solutions, which are connecting farmers to markets, reducing the transaction costs, enhancing traceability and transparency, and increasing the competitiveness of the sector.

    In 2018, there were 456 million mobile subscribers in sub-Saharan Africa, representing a subscriber penetration ratio of 44%. This number is projected to rise to about 600 million in 2025[17] and will further propel the emergence of innovations and foster greater adoption.

2. **Technological advances closing yield gaps:** There are innovations in seed systems driven by global and African research institutions, with improved seeds that are drought tolerant and can withstand pests and diseases, provide crop protection, and enhance soil health. These innovations, with appropriate commercialization, could generate a five- to twenty fold increase in many value chains. There are also advances in storage, processing, and distribution, which make it possible to bridge the gaps that exist between the supply and demand for food and increase the profitability for actors across priority value chains. These innovations will be explored in greater detail in Chapter 4.

3. **Increased youth engagement in agriculture**: 60% of Africa's 1.2 billion[18] people are under the age of 25. The World Bank predicts that by 2035, 350 million new jobs need to be created to sustain the growing population[19]. Despite more than 12 million youth entering the job market every year, only 25% will find formal employment. These youth are dynamic, creative, and energetic and can become either entrepreneurs or employees in the agriculture and food landscape[20]. Not surprisingly, the sector is starting to attract young entrepreneurs in the agtech, fintech, and digital landscapes, as well as food bloggers and chefs. As their businesses grow, they, in turn, will serve as critical champions and influencers, attracting more youth into the sector.
4. **The growing middle class and social media:** The growing middle class and the growth of social media have propelled the emergence of influencers in the food and chef industry and the celebration of African food, tradition, and culture. Today, several food and beverage festivals are bubbling up in African cities. Some examples are the Nairobi Food and Beer Festival, the Guaranty Trust Bank Food Fair, the Lagos International Beverage Fair, the DStv Delicious International Food & Music Festival, and Cape Town Street Food Festival. Africa has embraced this new wave of celebrating locally sourced food.
5. **Growing consciousness around diets:** With increased access to data and information, Africans are becoming more appreciative of the value of their indigenous food and the need to select the most appropriate foods and diets to sustain a healthy lifestyle. This growing consciousness is also fostering a backlash against fake foods and increased demand for fresh, organic produce with no added preservatives. Covid-19 has also reinforced the critical links between nutrition and health and the growing awareness that food is medicine. While the pace of their trend is relatively slow, it is noteworthy.
6. **Focus on equity:** The global protests against racism in 2020 have propelled many thought-leaders to assess the level of equity on the African continent and to demand greater accountability on many fronts. One glaring example that has generated significant concern and widespread calls for reform is in the African startup financing landscape, which invariably affects the agriculture and food landscapes. According to the *Guardian* UK article in July[21] 2020, of the top 10 African-based startups that received the highest amount of venture

capital in Africa last year, eight were led by foreigners. In Kenya, for instance, only 6% of startups that received more than $1m in 2019 were led by locals, a Viktoria Ventures analysis found. In Nigeria, 55% of the big money deals went to local founders and 56% for South Africa.

These realities are reflected on the ground, with many African entrepreneurs struggling to obtain funding.

Clearly, there is an urgent need to address the inherent biases and discrimination against African entrepreneurs in the food and agriculture landscape. There is also a sense of urgency for African leaders in the ecosystem to institute systems to ensure a level playing field and better support structures to equip entrepreneurs with the skills and tools to compete locally and globally. African fund managers and investors must provide catalytic financing and invest in entrepreneurs across the food and agriculture landscape. Finally, African policymakers must enforce laws that promote equity and support their local entrepreneurs.

Summary

Every entrepreneur who chooses to operate in the African agriculture and food landscape must understand the critical challenges that the sector currently faces and the increasingly negative impact of climate change. It is also imperative that African entrepreneurs appreciate the promising trends, such as the digital innovations, technological advances, youth engagement, growing consciousness around diets, and a deeper appreciation for African cuisine, which will enable scaling. These realities present a unique platform for entrepreneurs to enter and serve as catalysts for transforming the African agricultural landscape. Regardless of the value chain, the country, or the sector, opportunities abound.

Notes

1 Plant & Food Research. (n.d.) *Functional foods and ingredients for the premium consumer* [Online]. Available at: https://www.plantandfood.co.nz/growingfutures/food (Accessed: 10 August 2019).

2. Ezeh, A. and Tolu Feyissa, G. (2019). "What's driving Africa's population growth. And what can change it." *The Conservation*, 19 November [Online]. Available at: https://theconversation.com/whats-driving-africas-population-growth-and-what-can-change-it-126362 (Accessed: 30 August 2019).
3. Nugent, C. (2019) "The 10 countries most vulnerable to climate change will experience population booms in the coming decades," *Time*, 11 July [Online]. Available at: https://time.com/5621885/climate-change-population-growth/ (Accessed: 10 August 2020).
4. RELOAD Project. (n.d.) *Post harvest losses – A challenge for food security* [Online]. Available at: http://reload-globe.net/cms/index.php/research/7-post-harvest-losses-a-challenge-for-food-security (Accessed: 12 July 2020).
5. World Economic Forum. (2016) *Which countries spend the most on food? This map will show you* [Online]. Available at: https://www.weforum.org/agenda/2016/12/this-map-shows-how-much-each-country-spends-on-food/ (Accessed: 30 August 2019). Food and Agriculture Organization of the United Nations. (2015) *The economic lives of smallholder farmers: An analysis based on household data from nine countries* [Online]. Available at: http://www.fao.org/3/a-i5251e.pdf (Accessed: 30 August 2019).
6. UNICEF. (2019) *The state of the world's children 2019: Children, food and nutrition: Growing well in a changing world* [Online]. Available at: https://www.unicef.org/media/60826/file/SOWC-2019-EAP.pdf (Accessed: 24 August 2020).
7. UNICEF. (2019) *The state of the world's children 2019: Children, food and nutrition: Growing well in a changing world* [Online]. Available at: https://www.unicef.org/media/60826/file/SOWC-2019-EAP.pdf (Accessed: 24 August 2020).
8. Gouda, H.N., Charlson, F., Sorsdahl, K., Ahmadzada, S. Ferrari, A.J., Erskine, H., et al. (2019). "Burden of non-communicable diseases in sub-Saharan Africa, 1990–2017: Results from the *Global Burden of Disease Study 2017*," *The Lancet*, 7(10), E1375–E1387.
9. Food and Agriculture Organization of the United Nations. (2006) *Government of the Federal Republic of Nigeria: Support to NEPAD–CAADP implementation* [Online]. Available at: http://www.fao.org/3/ag063e/ag063e00.pdf (Accessed: 24 August 2020).
10. Sahel Consulting Agriculture and Nutrition. (2020) *Sahel quarterly: Talent in agriculture*, Pages 2–4 (23) [Online]. Available at: https://sahelconsult.com/talent-in-agriculture/ (Accessed: 28 August 2020).
11. Nigerian Forum for Agricultural Extension and Advisory Services. (2016) *Nigeria to employ 100,000 extension workers* [Online]. Available at: https://nifaas.org.ng/2016/09/01/nigeria-to-employ-100000-extension-workers/ (Accessed: 28 August 2020).

12 Smith, G.C. (2016) *What is food fraud?* [Online]. Available at: http://fsns.com/news/what-is-food-fraud (Accessed: 28 August 2020).

13 UN Women. (2019) *The gender gap in agricultural productivity in sub-Saharan Africa: Causes, costs and solutions* [Online]. Available at: https://www.unwomen.org/-/media/headquarters/attachments/sections/library/publications/2019/un-women-policy-brief-11-the-gender-gap-in-agricultural-productivity-in-sub-saharan-africa-en.pdf?la=en&vs=1943 (Accessed: 11 August 2020).

14 McKinsey & Company. (2020) *Safeguarding Africa's food systems through and beyond the crisis* [Online] Available at: https://www.mckinsey.com/featured-insights/middle-east-and-africa/safeguarding-africas-food-systems-through-and-beyond-the-crisis (Accessed: 11 August 2020).

15 Food and Agriculture Organization of the United Nations. (n.d.) *Commodities by region* [Online]. Available at: www.fao.org/faostat/en/#rankings/commodities_by_regions_exports (Accessed: 25 August 2019).

16 Tsan, M., Totapally, S., Hailu, M., & Addom, B. (2019) *The digitalisation of African agriculture report, 2018-2019* [Online]. Available at: https://www.cta.int/en/digitalisation/all/issue/the-digitalisation-of-african-agriculture-report-2018-2019-sid0d88610e2-d24e-4d6a-8257-455b43cf5ed6 (Accessed: 21 July 2020).

17 GSMA. 2019. *The mobile economy in sub-Saharan Africa* [Online]. Available at: https://www.gsmaintelligence.com/research/?file=36b5ca079193fa82332d09063d3595b5&download (Accessed: 29 August 2019).

18 Dews, F. (2019) "Charts of the week: Africa's changing demographics," *Brookings*, 18 January [Online]. Available at: https://www.brookings.edu/blog/brookings-now/2019/01/18/charts-of-the-week-africas-changing-demographics/ (Accessed: 11 August 2020).

19 Food and Agriculture Organization of the United Nations. (2018). "The future of Africa's agriculture rests with the youth," e-Agriculture, 13 June [Online]. Available at: http://www.fao.org/e-agriculture/news/future-africa%E2%80%99s-agriculture-rests-youth (Accessed: 30 August 2019).

20 Food and Agriculture Organization of the United Nations. (2018). "The future of Africa's agriculture rests with the youth," e-Agriculture, 13 June [Online]. Available at: http://www.fao.org/e-agriculture/news/future-africa%E2%80%99s-agriculture-rests-youth (Accessed: 30 August 2019).

21 Madowo, L. (2020) "Silicon Valley has deep pockets for African startups – If you're not African," *The Guardian*, 17 July [Online]. Available at: https://www.theguardian.com/business/2020/jul/17/african-businesses-black-entrepreneurs-us-investors (Accessed: 11 August 2020).

2

DEVELOPING A COMPELLING AND SUSTAINABLE BUSINESS MODEL

From the onset, be prepared to refine, modify, or even significantly alter your initial business model to ensure that your approach is demand-driven, profitable, and sustainable!

Introduction

When Mezuo Nwuneli and I decided to establish AACE Food Processing & Distribution Ltd. in 2009, we were propelled by a desire to address the high rates of postharvest losses and malnutrition in Nigeria and to introduce healthy nutritious food options sourced locally. Our first products were jams made from the best West African fruits. Many painful months into the process, after developing recipes, purchasing equipment, and obtaining regulatory approval, we realized that our business model would only be viable as a niche player, catering to the elite and expatriate population. The average Nigerian family did not consume jam as part of their breakfast meal, and the few who did would not place a premium on the use of real fruits versus imported alternatives, which were made of flavored water and sugar. In addition, securing glass jars, which was critical to ensure a long shelf life, made the price points unattractive for the country's masses. One year in, we had to decide between dramatically changing our business model or closing the company.

Our story is not unique. When Peter Njonjo and Grant Brooke of Twiga Foods decided to venture into the Kenyan agriculture landscape, they were first attracted by the opportunity to export bananas to the Middle East. They quickly realized that there was no structured supply chain or standardization in the sector and that if these issues were not resolved, their plans would fail. Similarly, when Nhlanhla Dlamini decided to leave his prestigious job at McKinsey to enter the South African agriculture landscape, he was excited about the opportunity to export premium meats from South Africa to the rest of the world. He soon was hit by the reality of the standards and the regulation blocks in the European and United States markets and had to pivot quickly.

All three stories have ended well, and AACE Foods, Twiga, and Maneli Foods are thriving in Nigeria, Kenya, and South Africa, respectively. Their founders had to adapt, refine, and in some cases, completely alter their business models to address the realities of the ecosystems in which they operated.

This chapter provides steps for developing a compelling business model in the African agriculture and food landscapes, as well as frameworks and insights for reducing the often circuitous and frustrating processes that many entrepreneurs face in this sector. It also provides practical examples that will inspire aspiring and emerging entrepreneurs to build compelling, profitable, and sustainable business models.

Finding your sweet spot

With the size and complexity of the agriculture and food landscapes and the diversity across the continent, there are countless opportunities for entrepreneurs to explore. However, determining an entry strategy and a product or service focus can be overwhelming. In addition, most entrepreneurs who enter the industry attempting to capitalize on an opportunity or solve a problem quickly realize that their initial hypothesis might have been flawed. Like peeling an onion, with every new insight or supplier and customer engagement, they quickly recognize that the root cause of the problem or the source of the opportunity leads them down a different path. While this reality is not unique to the agriculture sector, it is further complicated in this landscape, given the relative lack of data to guide and inform strategic decision making, the fragmented nature of the industry, and the silos under which most entrepreneurs operate.

There are a few ways to assess the sector when determining how to enter or grow your business. It would be best if you asked yourself a few questions:

What problem are you trying to solve, or what opportunity are you trying to capitalize upon? When establishing AACE Foods, we were compelled by the need to address the high rates of malnutrition in Nigeria, which, according to the DHS Survey[1] in 2008, revealed that 37% of the children under the age of 5 were stunted. We recognized that Nigeria was naturally endowed for agricultural excellence, given the abundance of arable land, rainfall, and indigenous varieties of crops that thrived. However, as a country, we had underinvested in local processing, which led to 40%–60% of our fruits and vegetables going to waste. We strongly believed that investing in local processing would enable us to provide nutritious food for our people, while at the same time addressing the issue of postharvest losses.

While we were drawn by the need to solve a problem, other entrepreneurs are attracted by the desire to make money. Regardless of what has propelled your initial motivation to enter the sector, it is imperative that you can clearly articulate your ingoing premise and support it with tangible data, where available. (Visit Nourishingafrica.com and browse through the data by country and value chain to identify gaps that need to be addressed.)

What is your primary value chain of interest? Unlike many world regions, almost anything can grow in most parts of Africa. It is the number one producer of cassava, yam, cocoa, cashew, sesame, and okra, and there are many entrepreneurs actively engaged in the fisheries, dairy, and poultry sectors[2]. However, determining which value chain to focus on is linked to your knowledge of the sector, passion for the value chain, and understanding of the opportunities available. It is important to realize that you select one value chain from the onset, and over time, as you understand more about the sector, you can include new value chains.

There are a few ways to prioritize value chains: your knowledge, interest, passion, perceived risk, required start-up capital, and time to break even. In addition, you have to determine whether your focus is on local versus international markets. For example, in Nigeria, export-focused value chains include cocoa, sesame, cashew, and hibiscus, while local market-focused chains include tomato, rice, maize, cassava, aquaculture, and poultry. In Kenya, export-focused chains include tea, cut flowers, and French beans versus maize, potatoes, and green leafy vegetables for food crops. Entrepreneurs are often attracted to the opportunity to generate foreign currency via exports. Still, they must realize that export markets require longer lead times for building relationships, securing contracts, and receiving payment.

Evans Danso, former Nestle Manager, recognized that there was an unmet demand for tilapia in his home country, Ghana. According to Evans, "demand for tilapia in Ghana is over 550,000 tons per year while production is less than 100,000 tons per year." To fill the gap, he started Flosell Farms Limited with a mission to supply an abundance of affordable protein for local markets and beyond. The company provides fast-growing male tilapia fingerlings to outgrowers and farms table-size fish in various sizes. Today, Flosell Farms[3] is the leading aquaculture company in Ghana and has also expanded to Zambia.

What role do you want to play within the value chain or ecosystem?
Beyond selecting your priority value chain, you must determine what activity you will undertake within that value chain. Attempting to engage in all aspects of the value chain from farm to fork from the onset is a recipe for early business failure.

The complexity and rigor associated with growing a business in each value chain demands significant market intelligence, resilience, and engagement with key stakeholders. For example, Nhlanhla Dlamini of Maneli Foods[4], pivoted to produce a range of nutritious pet food, after recognizing the numerous barriers to exporting premium meats for human consumption due to regulatory challenges and standards and that pet food was even more profitable. In addition to beef, lamb, and fish, the company leverages exotic sources of protein, including venison, crocodile, and even ostrich. In 2020, the company supplied a range of products to 12 countries across the globe.

If you are interested in processing maize flour, you must understand the entire maize value chain, including the type and source of seeds that farmers utilize to ensure standardization around issues such as the color of the maize and moisture content. It is also imperative that you understand the planting seasons, shelf life, postharvest management practices, aflatoxin levels, seasonality, quantity, quality, and pricing of the maize available in the local market versus international markets. In addition, you must understand the global market forces that affect the global maize industry, including trade dynamics that drive commodity price swings in the United States, China, and Brazil.

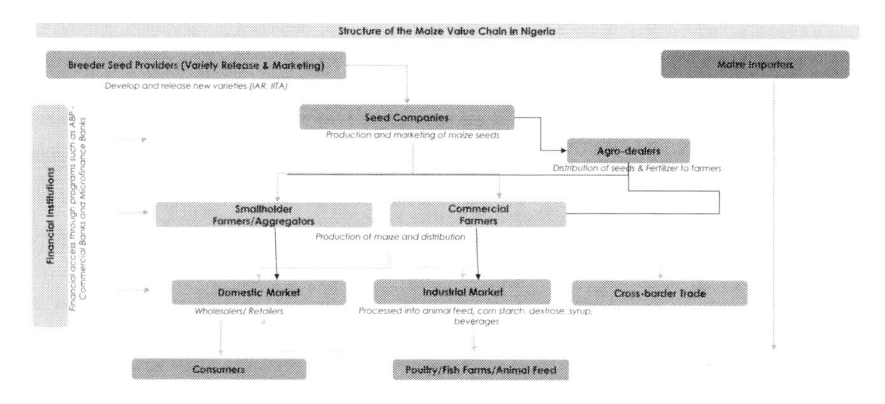

Figure 2.1 Sahel Consulting's maize value chain in Nigeria.

What is your primary region or country of interest? There are significant differences in the needs, opportunities, and infrastructure within countries and, indeed, across Africa. Consumer and customer profiles, the ability and willingness to pay, and the ease of doing business vary significantly. As a result, entrepreneurs must consider their knowledge and networks in a region or country, and the realities in that ecosystem when determining how to prioritize entry and establishment. For example, the Kewalram Chanrai Group established Olam Nigeria Plc[5], which started operations in Nigeria in 1989 and primarily focused on the exportation of cashews. The company started in the cashew belt and then gradually established a strong presence in the country. It then leveraged an "adjacency strategy," as an expansion tool. This strategy means that Olam has grown either by adding on new and related value chains within an existing country or by adding on the same value chain in another country. For example, from cashew in Nigeria, the company quickly added cotton, cocoa, and shea nuts to its list of exports, leveraging the same logistics and structures. It then expanded to Ghana by adding similar value chains. Today, Olam is a multibillion-dollar company, traded on the Singapore Stock Exchange, with operations in 70 countries.

Like Olam, companies engaged in the agriculture and food landscapes must determine where they will focus their efforts from the onset and then gradually decide how to expand.

WORKSHEET 2.1: MATRIX FOR FINDING YOUR SWEET SPOT

Value Chain/Entry Point	Inputs	Production	Storage/Logistics	Aggregation	Processing	Financing/Risk Management/ICT	Distribution	Retail	Food Service
Aquaculture/Fisheries									
Cereals and Grains									
Edible Oils									
Fruits and Nuts									
Insects									
Livestock and Small Ruminants									
Pulses									
Roots and Tubers									
Spices and Herbs									
Vegetables									
Others									

Who is your target customer?: Regardless of your geographic or value chain focus, the real crux of the value proposition of every business model is that it must add value or solve a problem for customers or clients. This essentially means that the business must be demand-driven and responsive to the needs of its target customer.

The reality in the agribusiness and food landscapes in Africa, given the relative complexity of the sector and the limited number of players, is that most entrepreneurs literally have to create the market – introducing the product or service to the customer/client and convincing them of the benefits of utilizing or consuming it. This is especially relevant for innovative products and technology tools that are new and untested and have to be introduced and gradually adopted.

There are multiple ways to segment customers/clients. One way is based on who will most benefit from the product or service. Typically, there are three ways to categorize this engagement:

- **Business to business:** This describes a business relationship between two companies, in which one company provides a product or service to another company. This is the model that Twiga[6] utilizes. The company realized that the fruit market was extremely fragmented and disorganized. There was variable quality, and farmers sold their fruits per unit of quantity and not by weight. There were high post-harvest losses estimated at 30%–40% of produce in Kenya[7]. As a result, the company decided to provide logistical and distribution services, serving as an intermediary between farmers and informal retailers. The cofounders, Peter Njonjo and Grant Brooke, leveraged technology to aggregate demand in retail and then used their buying power to amass supply. Leveraging this business model, Twiga basically ensures that informal retailers in Nairobi no longer have to wake up at 4 a.m. to go to the wholesale markets to purchase fruit, but, instead, can place orders utilizing their cell phones and then receive high-quality fruit within 14 hours.

  Like Twiga, AACE Foods recognized that many of the multinational noodle companies were importing their spice requirements either because they were unaware that the spices could be sourced locally, or were wary of their ability to identify consistent, high-quality raw materials. Over time, AACE Foods was able to convince these companies that it could meet international quality standards. For the past ten years, the company has processed spices and herbs in 25kg sacks for bulk customers who utilize them for their instant noodle spice sachets or as seasoning in other packaged food products. The company has expanded its institutional customer base to include the producers of snacks and biscuits, bouillon cubes, quick-service restaurants, and fast-food chains.

As reflected in the case of AACE Foods[8], there is further segmentation within the business-to-business classification based on the type of institutional customer you serve. Each of these customers has their own unique needs and specifications, and engaging with them requires further tweaking of product offerings and pack sizes to meet their needs. In addition, it is important to recognize that business-to-business sales are usually at lower margins because the product is being processed or repackaged for onward sale to the end consumer. As a result, suppliers are typically squeezed, especially if there is limited uniqueness and differentiation of the product or service offerings. In addition, the payment lead times and product visibility is usually limited. Finally, even though business-to-business partners often have long-term contracts, which allow for effective planning and scaling, the entrepreneur is often exposed to the inherent risks of depending on a few large customers, versus thousands or millions of smaller customers. Indeed, the successes and failures of their business model could either be a benefit or a burden. Entrepreneurs must carefully manage these relationships and hedge their risks to minimize exposure and dependence on one or two large customers.

- **Business to consumer:** This describes a business model in which a company provides its products and services directly to end customers and clients. For example, Maneli Pets offers a range of pet food products to customers in 12 countries. While a business-to-customer model often generates higher margins, demand is usually less predictable in the early years. Companies are required to build extensive distribution channels and a brand among consumers/customers to generate sizable and sustained demand. For food sales in particular, given that many countries still utilize open-air markets, building routes to markets and awareness among customers is expensive and takes an extended period. However, companies like Java Foods and Seba Foods, which produce nutritious food in Zambia, have proved that it is possible to build a strong business to consumer business, even in relatively smaller markets.

The process of building a brand and distribution network will be explored in detail in Chapter 5.

- **Business to government:** This describes the business model in which the company offers its services and products to the government. The Africa Improved Foods (AIF) company was established in December 2016 through a public-private partnership, described in detail in Chapter 7, produces and supplies Super Cereals. Its products are sold to the World Food Programme and the government of Rwanda.

   In the case of AIF, business-to-government sales have proven to be highly effective. However, this type of engagement is often risky, given the procurement processes of many African governments, the delays that providers sometimes experience with receiving payment, and the exposure to volatility in purchase agreements with regime changes. Engaging with the government typically requires more political connections and relationship management than the business-to-customer model but less advertising and branding requirements.

It is important to recognize that some companies straddle between the three customer segments, offering products or services to other businesses, consumers, and even governments. However, as is the case with many other sectors, entrepreneurs have to decide how to prioritize their customer focus, developing a business model that identifies their primary and secondary target customers. This can change over time as the business evolves and must be regularly assessed and tested. Many entrepreneurs often discover that the Pareto principle[9] – the 80:20 rule – applies in their companies, with 20% of their customers generating 80% of the sales or profits.

In addition, in some cases, a third party pays for the product or service for the end customer. This is a common practice for entrepreneurs who offer market information services or insurance to smallholder farmers. From the onset, many smallholder farmers cannot afford or may not value these services. In these cases, an international funder, financial institution, or even the government may pay for the products or services. Over time, as the product or service offering proves valuable, the farmers would be willing to pay directly, which alters the business model. The entrepreneur must remain agile and engaged to be able to respond to these changes in customer focus.

WORKSHEET 2.2: DETERMINING YOUR BUSINESS MODEL

- What problem are you trying to solve?
- What opportunity have you identified that you would like to capitalize upon?
- Who is your primary target customer?
  - Business to Business
  - Business to Consumer
  - Business to Government
- What value chain(s) will you focus on in the short- and medium-term?
- What aspect(s) of the value chain will you focus on?
- What region(s) will you focus on in the short and medium-term?
- How will you define success at scale?

Box 2.1: CONSIDER AACE FOODS TEAM'S RESPONSES TO SIMILAR QUESTIONS

What problem are you trying to solve?
The increasing rates of malnutrition, extreme dependency on imported food products. and high postharvest losses.

For who?
Smallholder farmers, food businesses, and consumers.

In what part of the value chain?
Herbs, vegetables, grains, and fruits – processing and distribution.

In what region?
Nigeria, the African continent, and the rest of the world.

How will you define success at scale?
Increased retail penetration across every Nigerian market and homes, closer collaborations with major multinationals and companies in the local and foreign food space, and increased growth in commodity export across the globe.

How will you define success at scale?
A nourished society dependent on homegrown solutions.

Articulating your mission, vision, and values

As you determine your business model, it is critical that you clearly articulate your company's mission, vision, and values. Many organizations misuse these terms interchangeably or even merge them. However, it is important to understand that each term is distinct and essential.

- A mission statement expresses the core purpose of an entity, clearly stating why it exists and setting clear boundaries around its principle activities.
- Values define the entity's enduring character, outlining key principles that are critical to the organization's existence for which it is unwilling to compromise.
- A vision statement provides a glimpse of success – the future desired state of achievement that the entrepreneur and his or her team are working to achieve.

While visions can be changed or altered, missions and values usually stay the same. Developing these critical statements is often an iterative process, and your key team and board members must be involved to ensure ownership.

Below are a few examples of statements of companies introduced in this chapter.

> **Box 2.2: EXAMPLES OF MISSION AND VISION STATEMENTS AND VALUES**
>
> **AACE Foods**[10]
>
> *Our Mission:* To provide nutritious and tasty food products from the best of West Africa's fruits, herbs, cereals, and vegetables.
>
> *Our Vision:* To be the preferred provider of food in West Africa, thereby contributing significantly to the improved nutritional status of our people, and better the livelihoods of our farmers.
>
> *Our Values:* Proudly West African, Quality Products, Cost & Time Efficiency, Continuous Improvement and Accountability.
>
> **Maneli Pets**[11]
>
> Our mission is to build Maneli Pets into the largest premium pet treat manufacturing company on the African continent and the preferred source of premium pet foods and treats for healthier, happier pets.

Olam[12]

Our vision: To be the most differentiated and valuable global food and agri-business by 2040.

Our purpose: To reimagine global agriculture and food systems. To support and double the production of food, feed and fiber (crop basis) to feed a growing world population estimated to be between 9.5 billion and 10 billion people by 2050 without destroying the planet and with less resources. It is also to transform the food system to produce more healthy food and reduce food wastage.

WORKSHEET 2.3: DEVELOPING OR REFINING YOUR MISSION, VISION, AND VALUES

Mission-specific questions:
- Why does your business exist?
- What is its core purpose?
- How will it achieve this core purpose?

Values-specific questions:
- What principles are critical for success?
- What principles are you unwilling to compromise with growth in reach and impact?

Vision-specific questions:
- What is your definition of success at scale?
- How will you know that you have arrived at your destination?

Business models that scale

Beyond finding your sweet spot or entry point into the agriculture and food landscape, you must determine if your business model is scalable and can achieve broader impact, revenue growth, profitability, and play a critical role that ensures that Africans can feed themselves and nourish the world.

Building on a scaling model, which is outlined in my early book, *Social Innovation in Africa: A Practical Guide for Scaling Impact*[13], enhanced with a singular focus on food and agriculture, there are at least six prerequisites for business models that scale in the African agriculture and food landscape as outlined below.

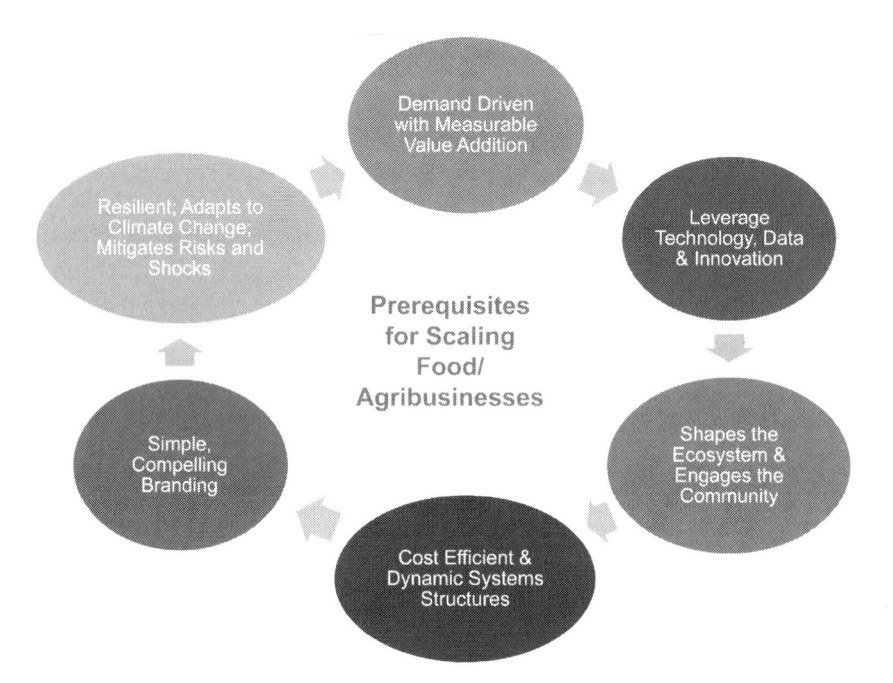

Figure 2.2 Prerequisites for business models that scale in the African agriculture and food landscape. Source: Adapted from a model shared in Ndidi Nwuneli's book Social Innovation in Africa: A Practical Guide for Scaling impact, Routledge 2016.

Demand-driven with measurable value addition: Any business model that is not demand-driven is not sustainable or scalable. However, it is often difficult to determine whether a business is demand-driven. Many agriculture and food innovations in the African context have been historically funded by international donors or propelled by government engagement. The sad reality is that most of these interventions were supply-driven, and not demand-driven, partly because they were subsidized solutions, and sometimes developed in research institutions or the West, far removed from the realities on the ground. They erroneously assume that the target population – often farmers – would be willing to utilize and even pay for the interventions to improve their lives. It is important to recognize that entrepreneurs who have already gone down the path of pushing supply-driven strategies have to be willing to alter their approach or target customer base to achieve high-impact results.

For true demand to be ascertained, you must address key aspects of demand-awareness, availability, accessibility, affordability, and acceptability. For example, improved seeds that could triple the yields of maize farmers in Northern Nigeria would only be used if the farmers can find them easily in their local market or local agro-dealers shops, if they are affordable, and if they look and feel like seeds – which is the acceptability component. However, proving that there is a sustainable demand for the seeds requires that the farmers use the seeds and achieve considerably higher yields, which convinces them to buy the seeds repeatedly.

Even in the Early Generation Seed System (EGSS) landscape, entrepreneurs are proving that they can build sustainable business models, despite donor and government distortions. Companies like Da-Algreen and Value Seeds in Nigeria offer hybrid maize seeds and a range of certified seeds, but they must invest in farmer education and awareness campaigns, complete with demonstration plots, in order to build a robust value chain.

WORKSHEET 2.4: ASSESSING WHETHER YOUR BUSINESS MODEL IS DEMAND-DRIVEN

- Is your business model demand-driven?
- Is this product or service needed? How do you know?
- Will enough people use it?
- Is any other group providing this intervention in your target community or region? (This will require a mapping of the market locally, nationally, and globally – to determine whether an adapted franchise approach might be preferred.)
- For a product – is it acceptable, accessible, and affordable, and is there considerable consumer awareness? If not, what would it take to educate consumers?
- What form of proof of concept have you utilized, and how has this induced trial and continued purchase?
- If your product/service disappeared today, would anyone miss it?

Measurable impact: Determining the impact of your business model can be exceedingly difficult and even expensive in the short-term but is critical for demonstrating value especially to potential customers, clients, or funders.

Most African entrepreneurs simply rely on sales and profitability data as a proxy for customer feedback and impact because of the scarcity of credible publicly available data on brands, market share, buying patterns and trends. Other entrepreneurs, who are willing to spend limited funds on measuring their impact, actively seek feedback through surveys, testimonials, and customer interviews. However, these two approaches, while often low-cost, can be misleading, and prove difficult to predict the future and provide insights into strategic decisions.

There are very few examples of private companies and social enterprises that actively measure the impact of their interventions. For example, the One Acre Fund[14], with operations in Burundi, Kenya, Malawi, Rwanda, Tanzania, and Uganda, benefits from considerable support from international foundations and development partners and is one organization in the agricultural landscape that actively evaluates historical data and plans for future years – not just in terms of scale, but also impact.

The One Acre Fund has an entire department called "Scale Innovations" that focuses on improving the scaling efforts' efficiency and effectiveness, guided by a key formula: Social Good = Reach × Impact. The team prioritizes four key objectives: understand the challenges, design program innovations, pilot the innovation, and scale up. Beyond reporting that it served more than 800,000 farmers in 2018, the organization states that it has helped 73% achieve financial sustainability and recorded a 97% farmer repayment rate, with the average farmer making $96 more in profits.

The emerging focus on baselines, control groups, and "theory of change" models – in which inputs, activities, outputs, and outcomes are measured – are all useful tools that the sector is starting to leverage. However, this trend is being stifled by the nature of the data collection practices and the quality of the data produced by many government agencies in Africa, which compels entrepreneurs to engage in the expensive process of primary data collection.

Once impact is measured, the results – whether positive or negative – should be reviewed internally to inform your strategic planning process and to enable you to make key decisions regarding resource allocations.

WORKSHEET 2.5: MEASURING IMPACT

- Is your business achieving measurable impact?
- What key indicators do you currently measure? How are they measured? What is the frequency and mode of data collection?
- What baseline or control group do you utilize to demonstrate a comparison between your intervention and the status quo? Is this credible? What does it demonstrate?
- How has measurement and evaluation of your innovation's impact informed your approach, changed your strategy, or influenced the behavior of key stakeholders?
- Has the intervention been tested for a long time to prove that it meets a need relative to alternatives? Is this need sustainable?
- What opportunities for improvement or failures have you identified, and how have they shaped your activities or informed your decision-making process?
- Have you utilized external evaluators to test and verify the impact of your interventions? What did this process reveal?
- How do you communicate your impact to internal and external customers, clients, funders, the government, and other stakeholders?

Summary

Defining your business model as an entrepreneur in the agriculture and food landscape is difficult because of the complexity and fragmentation of the landscape. However, you must take a proactive approach to define your scope from the onset – the problem that you are trying to address or opportunity that you hope to capitalize upon, your value chain, and your regional focus. You must define your mission, vision, and values and institute the systems and structures for ensuring a demand-driven and sustainable business model for scaling that is rooted in clear metrics for value addition. Finally, as reflected through the stories of entrepreneurs across Africa, be prepared to refine, modify, or even significantly alter your initial business model to ensure that your approach continues to be dynamic, innovative, profitable, sustainable, and withstand shocks.

Notes

1. The Demographic and Health Surveys Program. (2008) *Nigeria: Demographic and health survey 2008* [Online]. Available at: https://dhsprogram.com/publications/publication-FR222-DHS-Final-Reports.cfm. (Accessed: 22 July 2020).
2. Food and Agriculture Organization of the United Nations. *Nigeria at a glance* [Online]. Available at: http://www.fao.org/nigeria/fao-in-nigeria/nigeria-at-a-glance/en/ (Accessed: 22 July, 2020)
3. Flosell Farms. (n.d.) *Home page*. Available at: https://flosellfarms.business.site/ (Accessed: 24 August 2020).
4. Maneli Pets. (n.d.) *Home page* [Online]. Available at: http://manelipets.com/ (Accessed: 22 July 2020).
5. Africa Business Magazine. (2013) "Olam's blueprint for success," *Africa Business Magazine*, 13 August [Online]. Available at: https://africanbusinessmagazine.com/african-banker/olams-blueprint-for-success/#:~:text=Olam%20started%20business%20in%20Nigeria,we%20moved%20into%20other%20adjacencies. (Accessed: 22 August 2020).
6. On, D. (2019) "Twiga Foods secures $30M to digitize food distribution," Twiga Foods [Online]. Available at: https://twiga.ke/2019/10/28/twiga-foods-secures-30m-to-digitize-food-distribution/ (Accessed: 21 July 2020).
7. Sawicka, B. (2019) "Post-harvest losses of agricultural produce," in Leal Filho, W., Azul, A., Brandii, L., Özuyar, P., and Wall, T. (eds) *Zero Hunger: Encyclopedia of the UN Sustainable Development Goals*. Cham, Switzerland: Springer.
8. AACE Foods. *Home page* [Online]. Available at: https://aacefoods.com/ (Accessed: 22 July 2020).
9. Lavinsky, D. (2014) "Pareto Principle: How to use it to dramatically grow your business," Forbes, 20 January [Online]. Available at: https://www.forbes.com/sites/davelavinsky/2014/01/20/pareto-principle-how-to-use-it-to-dramatically-grow-your-business/#53f869ac3901 (Accessed: 10 November 2020).
10. AACE Foods Website; https://aacefoods.com/ (Accessed: 10 November 2020)
11. Maneli Pets Catalog; http://manelipets.com/media/Maneli-pets-catalog.pdf. Accessed: 10 November 2020).
12. Olam Website; https://www.olamgroup.com/investors/company-information/business-model-and-strategy.html. Accessed: 10 November 2020).
13. Nwuneli, N.O. (2016) *Social innovation in Africa: A practical guide for scaling impact*. Oxon, England: Routledge.
14. One Acre Fund. (2018). *Foundations for scale: 2018 annual report* [Online]. Available at: https://oneacrefund.org/2018-annual-report/ (Accessed: 22 July 2020).

3

TALENT FOR SCALING

Introduction

As reinforced by all of the key actors in the local and international debt and equity financial landscape interviewed for this book, the most important criteria for investment decisions in agriculture or food businesses is the passion, drive, and capacity of the entrepreneur and his/her leadership team. However, more than 70% of the entrepreneurs that formed part of our research cited finding, hiring, and retaining top talent as one of the most significant limitations to scaling their businesses. Clearly, there is a talent crisis in this sector.

These issues are not unique to the agriculture and food sector in the African context. The dearth of strong agribusiness and food processing training programs and the perception that agriculture is a poor man's profession have limited widespread interest by prospective students, with many viewing it as a last resort, or their third or fourth choice course options. Also, the historical focus on commodity exports dominated by large trading companies, limited investments, and the relatively low levels of value addition in the sector to date has further exacerbated the situation. As a result, many companies in this sector have struggled to attract and effectively retain and compensate the best talent.

The good news is that some entrepreneurs in the food and agriculture landscape have surmounted these obstacles by prioritizing talent engagement and management. They have enhanced their skills, built a strong leadership team, and constituted a strong board of directors. Also, they have determined which aspects of their operations they must address internally and what they can outsource and have invested in building strong organizational cultures.

Based on extensive research, case studies, and my personal experiences in the sector, this chapter provides the tools to build leadership and managerial skills, attract, and retain talent, establish a strong board of directors, engage volunteers and partners, and build a high-performance culture.

Building your capacity as a leader

As the founder, chief executive officer (CEO), or managing director of a food or agriculture business, you require some critical life and leadership skills to effectively lead a high-performing team and achieve results, especially in times of crises and uncertainty. In addition, the passion and vision that led you to start the business, while important, is not enough to sustain the growth of your business. Consider the list of skills below that you need to build or strengthen to lead your organization. It is essential to recognize that this list is not comprehensive or presented in order of importance.

Emotional intelligence[1]**:** The ability to understand, use, and manage your sentiments in constructive ways to communicate effectively, build trust with others, overcome challenges, handle stressful situations, and defuse conflict. It essentially means being aware that your emotions can drive your behavior and impact others (positively and negatively) and learning how to manage those emotions, both your own and others', especially when under pressure.

Resilience: The ability to cope with and rise to address the inevitable challenges, problems, and setbacks you will face in the course of your business life and emerge stronger from them.

Agility: The ability to assess and understand a situation quickly and respond efficiently and effectively, changing course as required.

Integrity: This refers to honest and truthful engagement with all stakeholders that is exemplified by a consistent and uncompromising adherence to strong moral and ethical principles and values.

Networking skills: The ability to establish, build, and sustain relationships personally and professionally. Networking skills are especially relevant as you work to scale your business and shape your ecosystem. It requires strong communication skills, including the art of active listening and asking questions as well as interpersonal skills.

Personal discipline: The ability to manage your time and resources effectively to prioritize wisely, work within set time constraints, and plan. As your business grows, you have to recognize the need for personal discipline, given the increased demands on your time during and after formal working hours. Your ability to create clear boundaries and controls in your life and your comfort level with saying no to people and initiatives that may derail you from accomplishing your goals. This is a skill that many entrepreneurs have to learn.

Strategic planning: This is the process of developing a roadmap for your business, from its current position to its desired goal. This is done by first assessing your current situation and developing the programmatic and operational requirements and steps to achieve your vision.

Financial planning and management: This consists of framing objectives, policies, procedures, programs, and budgets as they relate to the financial activities of your business. It includes determining your capital requirements, structuring financial policies, preparing budgets, monthly, quarterly, and annual financial statements, ratios, scenario planning, liquidity analysis, and engaging in internal and external audits.

Human resource management: This includes the process of recruiting, selecting, and inducting employees; providing orientation as well as training and development opportunities within their roles in the business; appraising their performance; deciding compensation; and providing benefits accordingly. Human resource management also includes maintaining proper relations with and motivating employees as well as ensuring employee safety, welfare, and health measures in compliance with labor laws.

Technology/Digital skills: The ability to safely, responsibly, creatively and effectively use the appropriate technology to communicate; access, collect, manage, integrate, and evaluate information and data; estimate and predict future needs, solve problems, and create solutions; build and share knowledge, improve learning, enhance the efficiency and effectiveness of your business; and liaise with suppliers, distributors, and customers. The COVID-19 epidemic revealed that every business leader must embrace technology and digital skills to sustain operations.

Building these skills takes time and requires a concerted effort. However, it would be best if you devoted the time needed through personal study, active trial and error, coaching, and mentoring to build the skills. In addition, embrace the habit of asking for feedback, not just from your friends and loved ones, but also from your team members. More specifically, adopt the practice of "upward" feedback as a critical component of your staff evaluation process to uncover any blind spots that you may have and enable you to build your leadership skills.

WORKSHEET 3.1: ASSESSING YOUR LEADERSHIP SKILLS

(Do not share your self-assessment or anyone's assessment with others to ensure unbiased responses. Always accept feedback as a gift.)

Skill (rate on a scale of 1 to 10, where 1 is low and 10 is high)	Self-Assessment	Assessment from a Personal Friend or Family Member	Assessment from a Mentor	Assessment from a Member of Your Team
Emotional Intelligence				
Resilience				
Agility				
Integrity				
Networking Skills				
Personal Discipline				
Strategic Planning				
Financial Planning and Management				
Human Resource Management				
Technology/Digital Skills				

After you have completed Worksheet 3:1, develop an action plan for building or strengthening the skills that generated the lowest ratings. Commit to some immediate actions, as well as medium- and long-term steps to fill these gaps. In addition, appoint one or two accountability partners to share your milestones and check your progress.

Establishing a strong board of directors

Beyond investing in yourself as the founder, the most important next step is to establish a strong board of directors. This critical step differentiates a one-person company from an enduring enterprise and enhances the credibility of the company with all local and international stakeholders.

As documented in LEAP Africa's *Get on Board: A Practical Guide to Building High Impact Board of Directors in Nigeria*,[2] a strong board has oversight and fiduciary responsibility by law; however, specifics of the roles and responsibilities of boards vary across Africa, based on national laws. In addition, functioning and high-impact boards provide advice and support, assist with fundraising efforts, evaluate the activities of the CEO/executive director (ED), and challenge the organization to aim high and achieve results. This group typically consists of legal, finance and accounting, branding and communications, and subject matter experts who understand the industry and can complement the entrepreneurs' skills and experiences.

Ideal boards are composed of between five and nine members, representing the gender, geographic, religious, and ethnic and diverse groups in the targeted customer demographics. These individuals must be willing to commit to actively participating in at least four board meetings a year and committee meetings, as required. The time commitments may vary depending on the company's life cycle and its needs, with new companies or those in crises requiring more frequent board meetings and more involvement.

Startups in the African context can manage expectations of their board members by requesting for a grace period from inception before having to pay annual and sitting fees. However, as your company grows, be prepared to offer sitting fees, annual fees, and other perks to your board members to sustain their interest and commitment.

Board members of nonprofits typically do not receive compensation for their services on the board; in fact, they are expected to provide financial and in-kind support to the organization. In return, they are made to feel valued and actively engaged in the organization's activities to gain a deeper appreciation for its work and impact.

It is also important that board members have clear term limits to allow for the infusion of "new blood" and energy. Ideally, they should serve for 3-year terms that are renewable for a second term if they remain committed and engaged. In addition, the board should engage in annual evaluations to measure their effectiveness and the contribution of individual members to the organization.

> **WORKSHEET 3.2: HOW EFFECTIVE IS YOUR BOARD?**
>
> - Do you have the most suitable board members for your company? Do they have the right set of skills, experiences, and values? Do they have passion and commitment? Are they willing to invest the time and energy required to enable your company to scale?
> - How frequently does your board meet?
> - Do you have a board charter that guides the board operations? Do you have a structured governance process for decision making?
> - Do you have functioning board committees?
> - How frequently do they meet?
> - Do you have a structured governance process for decision making?
> - Does your board have an annual evaluation process to measure its performance?
> - Does it formally assess the performance of the CEO/ED at least once a year?
> - How do you engage the board during periods of crisis?
> - What specific steps should you take to enhance the effectiveness of your board?

Consider the example of AACE Foods, which established a strong board from inception. As cofounders, Mezuo Nwuneli and I resisted the temptation to populate the board with friends. Instead, we actively identified individuals who we respected, who had the requisite skills and experiences to help us start and grow a successful company. These individuals also shared our values of integrity, hard work, and commitment, and agreed to devote their time, resources, and intellect to pushing the company to achieve results. Our pioneer board consisted of the following individuals, who brought key skills to the organization:

- Dr. U.K. Acholonu, founder of Bio-Organics Limited, a respected company that produced vitamin premixes for the food industry. Dr. Acholonu had a wealth of experiences and networks in the food and beverage manufacturing sector and strong contacts in the regulatory agencies.
- Dr. Fabian Ajogwu, a respected lawyer and senior advocate of Nigeria (SAN) who had a wealth of experience in corporate Nigeria, was the managing partner of Kenna Partners, a leading law firm and also the founder of the Society for Corporate Governance Nigeria (SCGN).
- Nwando Ajene, a respected marketing and brand executive with extensive experience in the financial services, experiential and digital marketing, and fast-moving consumer goods sectors.
- Patricia Umeron, a respected food inspector with more than 30 years of experience in the quality control, food technology, and analysis landscape in the United States.

This board represented a significant diversity of skills. The members' experiences complemented my strategy background and Mezuo's strong financial management skills, with years of investment banking, private equity, and corporate finance experiences in the United States, Nigeria, and Senegal. The board also represented age and gender diversity but, more importantly, was composed of passionate people, who leveraged their experiences, skills, and networks to support the company.

From the onset, we informed the board members that given AACE Foods' status as a startup, we could not afford to pay sitting or annual fees for the first 5 years of operations. Thankfully, this fact did not dissuade our prospective board members. Instead, they all gladly accepted to serve. We provided them with small end-of-year hampers, with the company's products and other personalized gifts to show our appreciation for their service.

Today, AACE provides very modest sitting fees, which some board members opt to donate into the staff revolving loan emergency fund, and we have continued to refresh the board with new members.

Engaging partners or cofounders

Entrepreneurs often consider the prospects of starting a business by themselves or entering into a partnership with two to three individuals who share their passion and values and offer complementary skills and have decided to work together to start and grow a business in the food and agriculture sector. Notably difficult, the key to successful partnerships is the founders' shared vision, values, willingness to split roles from the onset, and to stick to their core strengths.

There are many examples of failed partnerships in the food and agriculture landscape and some emerging success stories.

- Carl Jensen, Sunday Silungwe and Kellan Hays cofounded Good Nature Agro[3] in 2014, a Zambian for-profit social enterprise that believes in farmer-centric impact. All three cofounders bring different skills to the business. Carl serves as the CEO. A farmer, as well as an entrepreneurship and soil management graduate, he brings the technical know-how and understanding to the team. Sunday is a "self-confessed people's person" and has a natural flair for innovation. He works closely with partners and staff as the company's director of marketing and communications. Kellan is the board of advisors chairperson and advises on business development, processes and controls, fundraising, and partnerships.
- Nkiru Okpareke and Emeraba Tony-Uzoebo cofounded Enviro-Gro Farms Limited[4] in 2011. They also have two silent partners. Nkiru serves as the CEO and chief operating officer (COO) and is a trained engineer and a human resources (HR) expert who has worked with multinationals in various countries as a project manager and HR professional. She leverages these skills and experiences when managing personnel, field operations, and production on their farm. Nkiru also manages sales and marketing for the organization. Eme is the chief financial officer (CFO) and head of strategy and business development. A chartered accountant who has worked in financial planning and analysis in companies in various countries, she manages the company's finances and processes and drives improvements in business performance, revenue, and profitability. Eme is very strategic and commercially minded. Both Eme and Nkiru have MBAs from top North American universities.

- Joe Roques and Claudia Castellanos are founders of Black Mamba Foods,[5] a fair-trade manufacturer that produces a range of tasty sauces, pestos, chutneys, and jams from its factory located in Eswatini. Neighboring South Africa and the UK are some of its biggest markets, with other export destinations, including Germany, Australia, and Taiwan. The couple started Black Mamba out of their love for chilies and empowering the local women farmers in their community. Claudia serves as the managing director and the face of the business, while Joe is the creative director and oversees all the creative aspects of the brand.

The partnership experiences of agriculture and food entrepreneurs on the African continent reveal key criteria for selecting partners, similar to experiences shared by other entrepreneurs across the globe. Select individuals who:

- Share your values of hard work, integrity, and accountability;
- Share your vision and passion for the sector;
- Complement your skills and experience and fill specific skill gaps;
- Have a track record of success and credibility;
- Are willing to take on and share responsibilities;
- Are people you respect.

Attracting and retaining a strong talent base

Founders typically have the passion and skills to start but finding talent to fill the 2 to 20 person slots in a growing company is really difficult – finding great joiners is one of the biggest challenges facing founders. Ross Baird of Village Capital.[6]

The quote by Ross Baird, an impact investor who has supported many social innovators on the African continent, is echoed by others in the sector. Clearly, the recruitment process is a critical phase in building a dream team that can lead the scaling effort. The strategies that are utilized for finding talent vary, based on the positions, the nature of engagement, the level of skill and responsibility required, and compensation.

Every entrepreneur committed to scaling his or her agriculture and food business has to build a strong management cadre with the right values of humility, passion, and emotional intelligence. In terms of specific roles required to be filled in the company, different terms are used to describe these roles, depending on the size of the company and the nature of its operations. However, at least one of the job descriptions must focus on the accounting and financial management of the company, and another on ensuring operating efficiency and effectiveness. As the company grows, it will need marketing and sales personnel and HR and technical skills linked to the operations of the company.

Beyond key technical skills, these individuals must possess the right balance of business acumen, knowledge, communication, and interpersonal skills. Their values should also mirror those of the organization – rooted in a strong work ethic and integrity.

Box 3.1: TYPICAL ROLES IN AGRICULTURE AND FOOD BUSINESSES

Position	Responsibilities (may vary based on the size of the business and its history)
Chief Executive Officer (CEO)/ General Manager	• Sets strategy and direction. • Creates, lives, and breathes the company's culture, values, and behavior. • Hires and leads the company's executive team. • Designs and implements short- and long-term plans. • Makes critical managerial and operational decisions. • Communicates and manages relationships with the board of directors, investors, employees, customers, suppliers, and other key stakeholders.
Chief Financial Officer (CFO)/ Accounts/ Admin Manager	• Oversees the management of the company's finances and financial activities including payables, receivables, expenses, and treasury. • Advises the management on key business decisions based on financial analysis and projections. • Responsible for forecasting, cost-benefit analysis, cash flow management, and management of capital structure. • Responsible for the preparation of weekly, monthly, quarterly, and annual financial statements and filing all tax obligations.
Chief Technical Specialist (e.g. Chief Agronomist, Factory/Plant Managers)	• Leverages technical expertise to maximize productivity and boost efficiency in the operations of the business. • Organizes and monitors all technical activities of the business to achieve adequate yield/productivity in the operations of the business. • Oversees the daily operations of the business and organizes the team to deliver on goals and objectives. • Determines how the team uses technology to improve products and services.

Human Resources Manager	• Plans, implements, and evaluates employee relations to ensure a culture of excellence and productivity. • Sources and develops talent for the organization, which includes recruiting, interviewing, onboarding, and training, etc. • Maintains the work structure by updating job requirements and job descriptions for all positions. • Oversees training and support interventions for the team. • Manages employee benefits, performance appraisals, promotions, and exits.
Supply Chain/ Procurement	• Sources and transports inputs/raw materials and finished goods. • Develops and maintains various supply chain plans and strategies to optimize operational resources while executing cost reductions and inventory controls. • Promotes the design, development, and implementation of warehouse, distribution, and logistics solutions. • Forms productive partnerships and contracts to optimize supply chain cost, such as shipping lines, freight forwarders, customs house brokers, warehouse managers, and related third-party logical service providers.
Chief Sales/ Marketing Officer (CMO)/ Marketing and Branding Manager[a]	• Responsible for delivering the organization's revenue targets. • Creates and drives sales and brand strategies, such as positioning, scheduling, and determining the route-to-market strategy. • Identifies and analyzes relevant data to assess market trends and opportunities for acquiring customers across various categories and channels available to the company. • Creates and executes brand communication and marketing plans. • Manages new and existing customer relationships, ensuring that all customer issues and complaints are resolved and dealt with promptly to maintain trust. • Analyzes sales and profitability data to improve current and future sales, maximize profitability, and grow a well-diversified product portfolio. • Organizes and attends relevant trade and corporate events to promote the company and its products.

[a] This role is not necessarily stand-alone, can be merged with MD or COO roles depending on size and product of the organization

Recruitment: It is always wiser to hire slowly by engaging in an in-depth recruitment exercise, complete with interviews and assessment tests or activities linked to their future roles within the company. In addition, check references, credentials, payslips, and other documentation to verify employment and test for integrity.

Investing in a robust recruitment process is critical for identifying the most appropriate team members to join your organization. Sahel Consulting Agriculture & Nutrition, which operates two offices out of Lagos and Abuja, Nigeria, but works across West Africa, has developed a seven-step process for recruiting external hires. For one position advertised, the company receives between 300 and 1,000 resumes and has to go through the tedious screening process. However, the most challenging part of the engagement with potential hires is the reference checks. This has to include both formal and background checks to validate the credentials, work experience, and values of a prospective employee. It is not uncommon to find out through this process that potential hires lied about their work experience or roles, inflated their salaries, and stated that they were still employed, when they had been fired or laid off. Finding credible, independent references to validate the information that prospective candidates provide is often exceedingly difficult.

Attracting talent: The entrepreneurs who have been able to attract talent have utilized a range of strategies, including enhancing their image, leveraging stakeholder networks, using technology and recruitment agencies, and engaging the community.

Referrals are often cited as one of the most important avenues for attracting talent. As a result, entrepreneurs typically leverage their boards, staff, partners, and beneficiaries to serve as HR ambassadors and recruiters and encourage them to circulate clear job descriptions and generate significant interest in the organization. Some entrepreneurs even offer financial incentives to employees who refer capable individuals who eventually get hired.

As an entrepreneur, you have to be able to tell your story in a compelling way that will make people want to join the business and shape its history. Good talent will have to believe in your dream and vision to want to work with you. In addition, to attract mission-driven high achievers, you must actively develop and promote a positive reputation. This includes being recognized as high-impact, cutting-edge, and innovative. This reputation is actively promoted through social and traditional media and ultimately generates unsolicited interest from strong performers who are drawn to your culture and reputation and have a desire to have a meaningful impact in society.

To sustain any positive image in the public domain, there has to be significant substance. As a result, you must work diligently to create a transparent and accountable organization that has a strong reputation for ethics and impact. Ensure that your company also empowers its team members and builds ownership and strong relationships with key stakeholders in the public, private, and nonprofit sectors.

> **WORKSHEET 3.3: HOW EFFECTIVELY ARE YOU ATTRACTING STAR EMPLOYEES TO YOUR COMPANY?**
>
> - Have you clearly outlined why people should want to join your team and what makes your vision unique?
> - Have you clearly outlined the value proposition for them? What will they gain by joining your team?
> - Have you made your existing team members part of the image building for attracting talent?
> - How consistent is your message?
> - How authentic is your message?

SoFresh,[7] one of Nigeria's leading healthy food ready-to-eat chains selling a wide variety of fresh salads, smoothies, juices, parfait, and other quick, on-the-go healthy meals, was established by a couple – Abimbola and Olagoke Balogun in 2010 – and has developed a robust recruitment and hiring integration process. According to the founders, this is categorized into four steps: job analysis, screening, physical assessments, and robust onboarding.

Interestingly, SoFresh prefers to recruit in-house as opposed to externally, especially for their lower to midlevel positions. This is because they firmly believe that this not only creates a stronger foundation, but also engenders staff loyalty and pride. For entry-level employees, they use different advertising channels depending on the role. In 2020, they had ten outlets and more than 150 staff in Lagos and Abuja. Once hired, SoFresh has designed a 6-week onboarding program. The company's commitment to recruiting and retaining talent has contributed to its rapid growth.

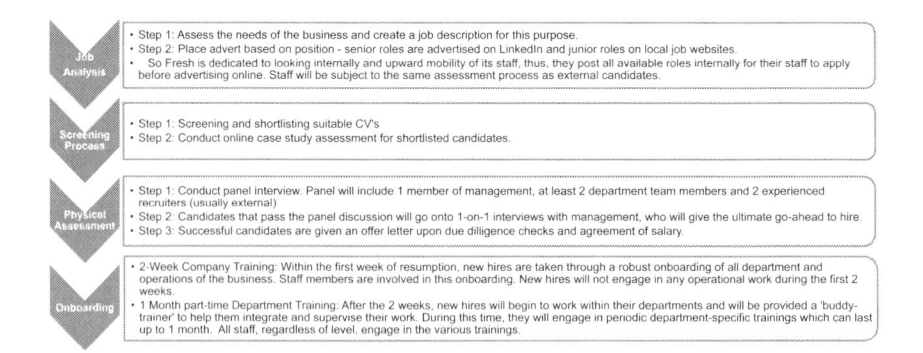

Job Analysis
- Step 1: Assess the needs of the business and create a job description for this purpose.
- Step 2: Place advert based on position - senior roles are advertised on LinkedIn and junior roles on local job websites.
  - So Fresh is dedicated to looking internally and upward mobility of its staff, thus, they post all available roles internally for their staff to apply before advertising online. Staff will be subject to the same assessment process as external candidates.

Screening Process
- Step 1: Screening and shortlisting suitable CV's
- Step 2: Conduct online case study assessment for shortlisted candidates.

Physical Assessment
- Step 1: Conduct panel interview. Panel will include 1 member of management, at least 2 department team members and 2 experienced recruiters (usually external)
- Step 2: Candidates that pass the panel discussion will go onto 1-on-1 interviews with management, who will give the ultimate go-ahead to hire.
- Step 3: Successful candidates are given an offer letter upon due diligence checks and agreement of salary.

Onboarding
- 2-Week Company Training: Within the first week of resumption, new hires are taken through a robust onboarding of all department and operations of the business. Staff members are involved in this onboarding. New hires will not engage in any operational work during the first 2 weeks.
- 1 Month part-time Department Training: After the 2 weeks, new hires will begin to work within their departments and will be provided a 'buddy-trainer' to help them integrate and supervise their work. During this time, they will engage in periodic department-specific trainings which can last up to 1 month. All staff, regardless of level, engage in the various trainings.

Figure 3.1 SoFresh's recruitment process. Source: SoFresh Nigeria Management.

Onboarding[8]: Onboarding includes the processes that allow new employees to learn about the organization, its structure, and its vision, mission, and values, as well as to complete an initial new-hire orientation process. Key steps to take when onboarding include:

- Contact new hires before their start date with important information and documents they should read before their first day.
- Create a robust onboarding curriculum including a welcome message from a member of management and an introduction to the company that includes an overview of business as well as the mission, vision, values and culture and a review of the Staff Handbook.
  - Induction by team members: Overview of work culture, organization structure, dress code, and performance evaluation process; overview of policies and procedures and the company's employee handbook; accounts and budget management onboarding.
  - Role onboarding: Clearly outline the roles and responsibilities of the new hire and the expectations from the new hire's manager and peers.
  - Soft- and hard-skills training.

The curriculum should be tailored to suit your business and the distinct function of the new hire while maintaining the essential elements. Onboarding should be a robust road map leading a new employee from orientation, through integration, and ultimately to full effectiveness. The length and depth of the program should differ by level and company structure. It is important to prioritize the employee experience, ensuring that you make the first day memorable. Providing a buddy/mentor to help new hires acclimate to the organization can help with this. Once the onboarding is complete, allow new hires to give feedback on the experience. A fresh perspective is always helpful to the company, and it also makes the new hires feel heard.

Organizational charts/structure: It is important that you invest in creating organizational charts, develop clear job descriptions, and review and refine them at least once a year. These charts will inform the hiring process, ensure organizational alignment, and enhance team culture.

In early-stage startups and other smaller firms, the founders may have to wear many hats and perform roles across different functions as business needs dictate. However, as the business grows, the need for clear job descriptions and division of roles and responsibilities is critical.

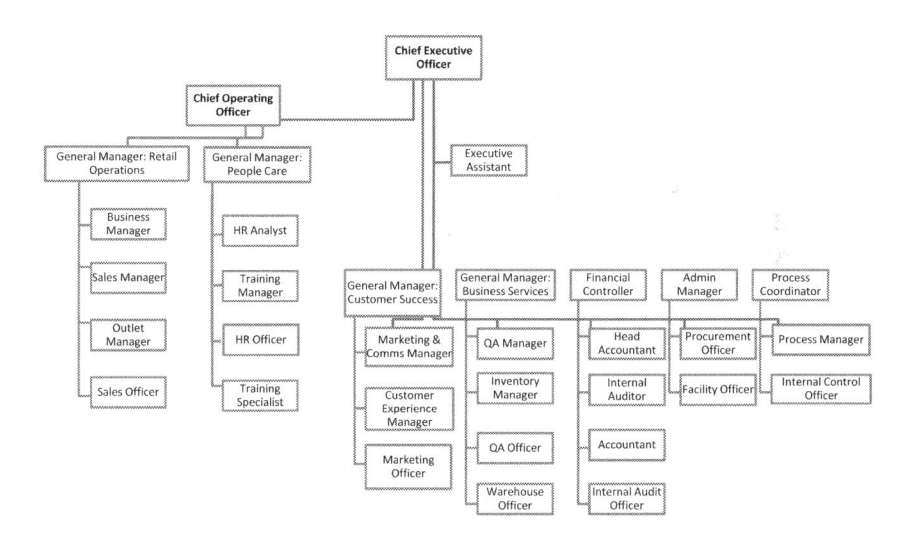

Figure 3.2 SoFresh 2020 organizational chart. Source: SoFresh Nigeria Management.

Volunteers

Agripreneurs, like many of their counterparts in other sectors, leverage volunteers for short- and long-term interventions. Some organizations even encourage individuals to volunteer as a pathway to paid employment. Others recognize this group as a pool of talent for training, facilitation, and support for key events or activities. The National Union of Coffee Agribusinesses and Farm Enterprises (NUCAFE),[9] led by Joseph Nkandu, employed 369 people in 2020. Fifty-three of these individuals were recognized as full-time employees, 251 as field/contract staff, and 65 as interns. NUCAFE extends internship opportunities to students from across the world to gain experience working for the organization. As of the end of 2019, NUCAFE's membership comprised of 213 farmer cooperatives/associations with 215,120 farming families and 1,512,210 individual coffee farmers in the five coffee-growing regions of Uganda.

Partners in Food Solutions aggregates and mobilizes teams of volunteers from its member companies – General Mills, Cargill, Royal DSM, Bühler, Hershey, and Ardent Mills. These volunteers transfer knowledge and technical assistance to processors in areas such as product formulation, fortification and packaging, and business operations. The companies also provide access to facilities and state-of-the-art equipment and infrastructure. Partners in Food Solutions has supported numerous small- and medium-size enterprises (SMEs) across Africa with enhancements to their factory operations, introduction of new products, purchase of equipment, and scaling. AACE Foods, Java Foods, and many of the other companies profiled in this book have benefited from the fantastic support provided by Partners in Food Solutions.

Fellows

Some international development agencies and funders organize Fellowship Programs through which they imbed technical talent within organizations for short periods to work on a specific task or strengthen the capacity of the company. Fellows are recruited from leading business schools or consulting firms. A few impact investors' innovations, such as Acumen and LGT Venture Philanthropy, operate fellows' programs. For example, the Acumen East Africa Fellows Program, a one-year, fully funded leadership program, enrolls 20 fellows from across East Africa annually. The fellows are provided with five week-long seminars focused on building their financial and operational skills and connecting them to the local and global community. International fellows often benefit from free housing, transportation, and a stipend. This cost is typically borne by the agribusiness, the impact investor, or shared by both and is often prohibitive for smaller, local organizations.

Short-term consultants

When agriculture and food entrepreneurs require targeted expertise and interventions in key areas of work, they also engage short-term consultants. Global best practice stipulates that you design a formal Request for Proposal (RFP) process to identify the most suitable consultant. Given the significant variability of skills and capabilities of short-term consultants, always assess a minimum of three proposals to gauge the suitability of the proposed approach, the proposed budget, and the work plan. Always check references to ensure that you engage individuals who have a positive track record and significant expertise. Formal contracts are critical to ensure that both the consultants' and the organization's expectations are aligned, with clear timelines for deliverables and payments linked to clear milestones and results. Even when a third-party or funder is paying for the services of a consultant, always insist on a due diligence process to ensure that the selected candidate is well suited and can add the most value.

It is important to recognize that sometimes volunteers, fellows, and consultants may prove to be more expensive than full-time hires, primarily if they are recruited from a different city and require housing and transportation. These costs, in many African cities, prove prohibitive and sometimes detract from the cost savings that organizations expect to generate from engaging them in the first place.

- **Locals/Expatriates:** Entrepreneurs often struggle to find local talent to fill specific gaps in their operations. In many cases, they have resorted to engaging staff from Asia or Europe who have functional expertise but are often more expensive and require work permits and, in some cases, housing, health insurance, and other perks. Consider the example of Java Foods in Zambia, which in 2019 had 26 full-time employees and recruited a significant number of expatriates from Kenya and Uganda for higher-level positions in critical areas including production, quality control, and maintenance.

If you must engage expatriates, ensure that they have the expertise and experience to add significant value to your organization and help it scale. More importantly, ensure that they invest in training the local staff who will eventually take over their roles, with clear timelines for the succession process.

Outsourcing

Founders and management teams of agriculture and food businesses must recognize that some aspects of their operations can be outsourced to other entities or individuals who are better positioned to manage them. The component should be determined based on the company's products, services, or processes. For example, while some organizations outsource their accounting and finance roles or legal services, others outsource their marketing, sales, and measurement and evaluation functions. Some entrepreneurs in the food landscape even outsource production, leveraging contract manufacturing services provided by other companies.

Many companies in the food and agriculture landscape also have to engage contract staff or workers – individuals who are not directly on payroll but provide support services. These range from serving as contract growers for seeds, to selling the seeds and fertilizer, to merchandising products in the organized formal channel or serving as sales executives in the informal retail channels.

> **Box 3.2: IITA GOSEED – OUTSOURCING**
>
> IITA GoSeed, a startup seed company incubated by International Institute for Tropical Agriculture (IITA) Ibadan, engages outgrowers for foundation seed production, whereby the contract growers are "recruited" based on set criteria, such as having several years' experience producing the crop, a specific size of farmland, maintaining good standing in the farming community, etc. In addition, IITA GoSeed Ltd. onboards new outgrowers using field demonstrations to show required management practices, sharing well-structured standard operating procedures with the outgrower, and facilitating follow-up training at intervals along the production cycle. The outgrowers are also required to sign a memorandum of understanding (MOU) that documents established terms, such as seed buy-back price, duration of engagement, standard production practices,

etc. Using this model, IITA GoSeed needs to keep only a few agronomists and one operations manager on its payroll who manage the outgrowers by helping to build their capacities and ensuring quality control, thus ensuring a far wider coverage that was not possible otherwise. This also minimizes production costs that would have otherwise been spent acquiring and managing large expanses of fields across the country.

IITA GoSeed also contributes to the effectiveness of the cassava seed value chain by building capacities of independent foundation seed producers. The outgrowers engaged are often graduated to independent foundation seed producers based on their experiences from producing for IITA GoSeed. The outgrowers are nurtured in technical and business management of seed production. They purchase and pay for breeder seeds from IITA GoSeed and make their own technical and business decisions with guidance from IITA GoSeed. Their registration with the regulatory body responsible for seed production in Nigeria (NASC) is facilitated by IITA GoSeed. The idea of having these independent foundation seed producers is (i) to empower farmers in the cassava seed sector, (ii) decentralize the production of early generation seeds of cassava with traceable origins, and (iii) meet the demand of cassava seeds with respect to specific varieties. In all of these, IITA GoSeed remains a core seed producer of cassava stems of new varieties developed by the IITA cassava breeding unit.

You need to weigh your organizational needs against your available budget to determine whether to hire full-time employees, engage volunteers, fellows, short-term consultants, or even outsource specific functions. Ultimately, your goal is to minimize your overhead costs while leveraging talent and expertise in the community or from across the globe to drive high-impact scaling. Managing individuals who are not on the payroll and who work in the field is often difficult unless activities are actively monitored and incentives are linked to performance.

Building a high-performance and conductive organizational culture

Human resources: All businesses need a clear HR strategy as a pivotal component of their operating model. Sadly, this critical component of operations is often ignored until there is a crisis.

For smaller businesses with tight finances, having an HR person/department may be difficult. However, establishing a small and efficient HR team, even if it means only one person to start, is critical to growing your business, building your company culture, and managing employees. HR personnel should be able to multitask, have high-energy, and the ability to switch gears quickly. As the company grows, the HR role will evolve, so it is important to have an agile person who can adapt to a changing environment.

Responsibilities of HR can be categorized into six main functions:

1. Recruitment and talent management – Direct all hiring and onboarding procedures for new employees. This includes advertising vacancies, shortlisting eligible candidates, salary determination, position classification, and grading scale of employees.
2. Workplace safety – Continuously educate and train employees on company policies as well as keep your employee handbook up to date. Promote a positive and open work environment. Monitor employee progress and monitor your company climate and culture to ensure that it remains productive and employees are in positive states.
3. Employee relations – Work with management to strategically plan HR initiatives that will benefit the company and encourage more efficient and beneficial work from employees. Facilitate cross-functional engagement between management and employees. Conduct regular meetings with employees for progress reviews, assessments, and to discuss any problems or grievances they may have. Coordinate and direct work activities for employees, and work with management to assess, reassign, promote, and/or release staff.
4. Compensation planning – Manage employee benefits, health plans, pensions/retirement plans, etc.
5. Labor law compliance – Understand and adhere to all labor laws.
6. Training – Coordinate and/or provide information on regular training opportunities for staff.

In addition, HR professionals need a strong understanding of technology, such as automated sourcing tools, talent acquisition software, and candidate relationship management integrations. Competent HR personnel should know how to implement technology to reduce the amount of time spent on administrative tasks, improve data quality, and free up more time for value-added tasks.

Communications: There are critical systems and structures for building a strong and efficient culture. At the very least, there has to be a system for information to flow within the organization, even if it operates in multiple sites. This includes the use of email, WhatsApp, Slack, Zoom calls, town halls, memos, and other robust communication mechanisms. Communication has to be frequent and consistent.

For example, Sahel Consulting Agriculture & Nutrition Ltd. has weekly meetings every Monday at 9 a.m., in which all team members, including HR and support staff, provide briefings on tasks accomplished in the previous week and their plans for the upcoming week. Sahel also has project kick-off and closing meetings that are facilitated by the team leaders. Through these meetings, team members are introduced to their projects, key performance indicators (KPIs) are developed, and expectations are discussed. Toward the end of the project, the team also engages in a series of meetings in which KPIs are evaluated. The HR team also conducts a peer-review process, and project documents are reviewed against proposed objectives.

Staff salaries, bonuses, promotions, and other incentives

As an entrepreneur, you have to offer a bundle of incentives to your employees to attract and retain them. Consider the following.

- **Compensation:** If you want to keep top-notch talent, then you will have to pay them well. Salaries are typically determined based on an employee's skill and experience, supply and demand of talent, geographical location, and worker seniority. It is critical that you define an employee pay scale and that this forms a vital part of the staff induction. For consultants and contract employees, payment should be linked to the contract or prestated payment structure based on daily rates or milestones achieved. For interns and volunteers, the best practice is to pay stipends that covers transportation and some costs.
- **Benefits and bonuses:** There are a range of benefits that employees deserve and others that will keep them engaged and committed. The statutory benefits such as pensions are linked to country policies and the size of the company. Others are related to the company's philosophy and values. Consider providing some of the following benefits to your team members.

- **Life and health insurance:** It is important to provide your staff with some form of life and health coverage, as this will prove more cost-effective than paying for costs as they arise when a team member is ill or dies unexpectedly. The type of insurance that is selected and coverage benefits provided can be linked to the seniority of the employee, his/her role in the company, and needs. Also, consider including the children and spouse of the employee on the health insurance plan. It is important to note that many health maintenance organizations (HMOs) allow for semiannual payments and the switching of employee names, which reduces the risks and costs associated with employees exiting early.
- **13th month salary and performance bonuses:** Some companies pay all confirmed staff an extra month's salary at the end of the year. Others chose to link payouts to the financial performance of the company and the results of the employees' evaluation. Some companies provide bonuses equivalent to an annual salary. For example, Good Nature Agro rewards its Private Extension Agents (PEAs) with a percentage of company revenue that is determined entirely by the quantity and quality of seed produced by the farmers they manage. Others do not pay bonuses at all.
- **Equity:** Equity plans could be offered to full-time employees, advisors, and board members. Offering equity allows startups to compete financially for the talent they need. This also serves as an incentive for staff to remain in the organization and builds ownership, as staff would be required to commit to work for a stipulated period before their shares can be vested.
- **Leave:** A company must provide paid annual leave to all staff members. Best practice stipulates 15–20 days per year. A common pitfall occurs when management and/or work culture discourages employees from taking annual leave. Employees should also benefit from sick leave, examination leave, personal days, public holidays, and compassionate leave. Maternity and paternity leave should be offered in line with government regulations, at the barest minimum.
- **Training:** Life-long learning should be a core value of any company, and staff must engage in continuous learning and development as part of the company's roadmap to success. Training and development includes engaging experts to conduct trainings within your organization, enrolling your team members in external programs or

on-the-job training via peer-learning activities. Regardless of mechanism, introduce training plans for team members, clear criteria for selecting programs and measuring impact, and a mechanism for cascading the insights and skills gained to the rest of the organization.

There are so many free programs in the sector, and management should prioritize employee training and encourage staff to leverage them. In addition, as an entrepreneur, you should balance internal and external training, inviting qualified professionals to speak to your team members and actively nominating employees, in a transparent manner, for free and paid training programs offered locally, nationally, regionally, or internationally.

In addition, Sahel occasionally engages experts to train employees in-house and bears the full cost of these trainings. Sahel also has weekly team events, including knowledge worth-sharing sessions in which each team member takes turns to teach or present relevant practices or current trends to the rest of the team, facilitating peer-learning within the company.

According to Evans Danso of Flossell Farms, through the company's partnership with the University of Alabama, his staff benefit from training in financial literacy, new technologies, and global best practices. These programs are offered by Ph.D. and master's students who visit his farms in Ghana for 10–14 days as part of their course requirements. In turn, the students are provided with practical material for their theses and dissertations.

Recognition and motivation by management: As the CEO, you must continuously motivate and reaffirm your team members to keep company morale high. Show recognition and express gratitude when deserved. Your employees should be aware of their impact on your growth and success. Consider introducing "Employee of the Month or Year," awards, end of year prizes, and peer-recognition initiatives. Include cash or in-kind gifts, and nominate your team members for national and international awards, programs, and fellowships.

Performance evaluations

Your employees must understand how their performance will be measured and evaluated from the orientation program. They should understand the process, the key performance indicators that will be assessed, and the frequency of the assessment. Also, encourage peer-based feedback as part of the evaluation process. You must ensure that the evaluation process is transparent and perceived as fair.

If your business requires employees to work in relatively fixed teams, consider conducting team evaluations in addition to individual performance evaluations. This includes assessing team goals and objectives, peer assessment and feedback, evaluation of the performance culture within the team, key learning, and improvement opportunities.

You should provide your team members with explicit feedback on their performance on the job and contributions to the organization's success. Their ability to uphold the company's values of integrity, personal sacrifices, and sense of ownership should be assessed and recognized. In addition, you should work with them to explore their personal goals and career path and help them to understand how their performance affects their progress. The evaluations should also include a work plan for training, coaching, and mentoring.

Good Nature Agro engages their PEAs on a contract basis. The company invests in a robust 3-week training program, called the PEA College, to train these agents. It also has a detailed manual and clear KPIs, with supporting systems and structures.

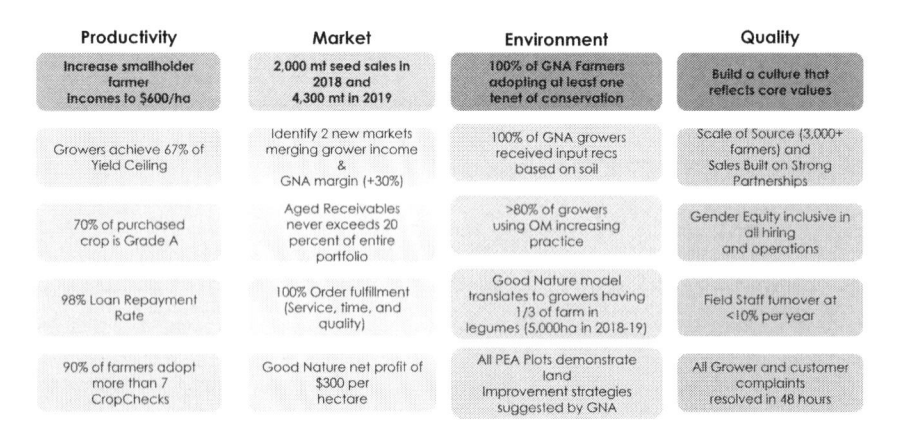

Productivity	Market	Environment	Quality
Increase smallholder farmer incomes to $600/ha	2,000 mt seed sales in 2018 and 4,300 mt in 2019	100% of GNA Farmers adopting at least one tenet of conservation	Build a culture that reflects core values
Growers achieve 67% of Yield Ceiling	Identify 2 new markets merging grower income & GNA margin (+30%)	100% of GNA growers received input recs based on soil	Scale of Source (3,000+ farmers) and Sales Built on Strong Partnerships
70% of purchased crop is Grade A	Aged Receivables never exceeds 20 percent of entire portfolio	>80% of growers using OM increasing practice	Gender Equity inclusive in all hiring and operations
98% Loan Repayment Rate	100% Order fulfillment (Service, time, and quality)	Good Nature model translates to growers having 1/3 of farm in legumes (5,000ha in 2018-19)	Field Staff turnover at <10% per year
90% of farmers adopt more than 7 CropChecks	Good Nature net profit of $300 per hectare	All PEA Plots demonstrate land Improvement strategies suggested by GNA	All Grower and customer complaints resolved in 48 hours

Figure 3.3 Performance measurement indicators utilized by Good Nature Agro, Zambia. Source: Good Nature Agro Management.

Exiting low-performing staff: Talent management includes the exiting of staff whose values and/or performance does not match the expectations set by the leadership. In cases in which there was a hiring mistake or a simple mismatch of skills and values, a quick layoff, once identified, is important.

However, it is often critical to give employees some time, leading up to the confirmation period, which is usually at the 6-month mark to determine whether there was a mismatch and a need for an exit, or whether the employee can learn, change, and eventually add value. If your organization is large enough, you may also consider moving them into a new department or new role, if they demonstrate that they may be more suited to a different position.

A few red flags that may reveal that an employee should be exited include if the employee continuously struggles to fit into the organization, he/she is disengaged from work and projects, he/she makes frequent excuses for underperforming, or he/she has been provided with all the necessary tools to perform yet is not doing so.

During my time at McKinsey, I learned about a simple way to evaluate staff called the skill-will matrix.[10] It will help you assess your team members based on their commitment to learning and growing and their abilities.

WORKSHEET 3.4: MAPPING YOUR TEAM

	High Skill	*Low Skill*
High Will	**Contributors** (Empower these team members to take on more leadership roles in the organization) 1. 2. 3. 4. 5. 6.	**High Performers** (Provide training support to build their skills and challenge them to step up) 1. 2. 3. 4. 5. 6.
Low Will	**Low Performers** (You may want to try micromanaging them for a predefined period by putting them on a strict performance improvement and training plan. If this does not work, then you have to exit them from the organization.) 1. 2. 3. 4. 5. 6.	**Potential Detractors** (You may want to put them on a strict performance plan, and also introduce incentives to motivate them) 1. 2. 3. 4. 5. 6.

Building a strong company culture

Experience from entrepreneurs across Africa reveals that there are several benefits to creating and maintaining a healthy and positive work culture. Whether or not you are explicitly creating a culture in your company, one is emerging. As a result, it is imperative that you proactively create one that aligns with your values and vision.

The culture drives the way employees relate to one another, the way employees approach the work they do, and how employees relate with the company. Therefore, a company's work culture should not be left to chance; it needs to grow out of an intentional set of values and must be supported by a strong HR department.

A transparent culture ensures that employee sentiments can be understood, improves retention and recruitment, and allows employees to chart their career development within an organization. This also demands that complaints are handled efficiently and transparently. This requires the establishment of a system in which responses are quick, fair, and confidential. In addition, outcomes need to be driven by the facts. Apart from enabling HR departments to process complaints quickly, this also assures employees that their well-being is a priority and that their complaints are carefully considered.

A learning culture ensures that team members prioritize personal and professional development. They read widely and actively engage in training programs. At Sahel Consulting, I encourage team members to write articles and speak at local and international conferences to ensure that their voices are amplified in the public domain, and they receive recognition for their work.

A culture that promotes diversity and inclusion ensures respect and zero tolerance for discrimination and sexual harassment. A culture that fosters integrity promotes transparency and zero tolerance for unethical behavior. These attributes are also actively monitored in performance evaluations to ensure that team members are rewarded for the same values as well.

You should also work diligently to build a cohesive team by promoting ownership and open communication and ensuring regular communication, cooperation, and collaboration. Weekly staff meetings, annual staff retreats, monthly happy hours, end-of-year parties, and other vital mechanisms can play a role in building a strong team culture.

In addition, celebrate weddings, births, graduations, and other milestones, and also mourn with colleagues who have lost a family member. Your companies should develop clear policies to guide the formal contribution to these expenses, as well as employee contributions, as required to ensure consistent engagement. Work with your HR associate to design interventions that teach collaboration and empathy and enable team members to gain comfort with collaborative feedback. You must reward teamwork and not individualism by incorporating this quality as a critical component of employee evaluation.

Working with Millennials: Research conducted by Gallup[11] reveals that millennials have a different approach to workplace productivity and flexibility. This generation does not believe that productivity should be measured by the number of hours worked at the office, but rather by the output of the work performed. They are motivated by purpose, a strong desire to develop themselves, and feel valued and appreciated. Instead of bosses, they want coaches, and they are more interested in leveraging their strengths than addressing their weaknesses. Instead of evaluations, they prefer ongoing conversations.

Using technology, the function of remote/flexible working has become extremely popular for top talent who wish to work in cutting-edge organizations. Options millennials seek:

- Telecommuting – The opportunity to work from home or a convenient, remote location as needed.
- Part-time schedule: A scenario in which an employee works fewer set hours or is allowed to select different hours each week.
- Flexible schedule: The employee may work full- or part-time, but the hours are based on core times and flexible time bands.
- Alternative schedule: Working outside the typical 9-to-5 schedule, typically to accommodate other responsibilities.

Summary

Entrepreneurs committed to scaling in the agribusiness and food landscapes must invest in their personal development as leaders and innovators and attract, motivate, and retain a capable team of diverse individuals who are humble, motivated, and emotionally intelligent. They must also engage committed partners, boards of directors, leverage volunteers and consultants, and outsource key functions, where appropriate. Finally, they must invest in building a culture of excellence, integrity, and innovation that is guided by clear systems and structures and a strong performance management culture.

Notes

1. Segal, J. Smith, M., Robinson, L., & Shubin, J. (2019) *Improving emotional intelligence (EQ)*. Available at: https://www.helpguide.org/articles/mental-health/emotional-intelligence-eq.htm#:~:text=Emotional%20intelligence%20(otherwise%20known%20as,overcome%20challenges%20and%20defuse%20conflict (Accessed: 28 August 2020).
2. LEAP Africa. *Get on Board: A Practical Guide to Building High Impact Boards of Directors in Nigeria;* Farafina 2007
3. Good Nature Agro. (n.d.) *Our team* [Online]. Available at: https://goodnatureagro.com/team/ (Accessed: 25 July 2020).
4. Enviro-Gro Farms. (n.d.) *Our team* [Online]. Available at: https://www.envirogrofarms.com/our-team (Accessed: 25 July 2020).
5. Black Mamba Chilli. (n.d.) *Home page* [Online]. Available at: http://blackmambachilli.com/ (Accessed: 25 July 2020).
6. Nwuneli, N.O. (2016) *Social innovation in Africa: A practical guide for scaling impact.* Routledge.
7. SoFresh. (n.d.) *SoFresh locations* [Online]. Available at: https://sofreshng.com/locations (Accessed: 25 July 2020).
8. Hollister, R. and Watkins, M.D. (2019) "How to onboard new hires at every level," *Harvard Business Review.* 27 June [Online]. Available at: https://hbr.org/2019/06/how-to-onboard-new-hires-at-every-level (Accessed: 25 July 2020).
9. National Union of Coffee Agribusinesses and Farm Enterprises (NUCAFE). (n.d.) *About NUCAFE* [Online]. Available at: https://www.nucafe.org/index.php/about-nucafe/who-we-are (Accessed: 25 July 2020).
10. Stratechi. *Will skill matrix* [Online]. Available at: https://www.stratechi.com/will-skill-matrix/ (Accessed: 25 July 2020).
11. Gallup. (2016) *How Millennials want to work and live* [Online]. Available at: https://www.gallup.com/workplace/238073/millennials-work-live.aspx (Accessed: 26 August 2020).

4

LEVERAGING TECHNOLOGY AND INNOVATION TO SCALE YOUR BUSINESS

Introduction

In May 2020, a Forbes[1] article shared a simple question, "Who led the digital transformation of your company? – the CEO, CTO or COVID-19?" The reality is that most entrepreneurs would choose COVID-19 as their response to this question. With lockdowns and movement restrictions, many companies that had not previously invested in technology and innovation struggled for survival, and some even shut down. The COVID-19 pandemic exposed the inherent technology and innovation gaps in business operations and forced entrepreneurs to recognize the critical importance of embracing digital innovation and technology to survive and thrive.

This chapter makes a strong case for why entrepreneurs committed to scaling their agribusinesses must harness the power of technology and innovation to improve their product and service delivery, expand their coverage, increase their brand visibility, reduce their costs, and enhance their resilience. It also addresses how technology is transforming each stage of the agriculture and food ecosystem in Africa and how entrepreneurs can leverage these interventions in their businesses to scale. Finally, it explores strategies for digital and agtech entrepreneurs to enhance their business operations and drive growth, profitability, and impact.

Digitalization for agriculture

As defined by the Technical Centre for Agricultural and Rural Cooperation (CTA)[2], the term *digitalization for agriculture* is the use of digital technologies, innovations, and data to transform business models and practices across the agriculture value chain, including production, postharvest handling, market access, finance, and supply chain management.

A 2018 study conducted by CTA and Dalberg[3] identified 390 active digital innovations on the African continent, with a focus spanning from financial inclusion, advisory access, market linkages, supply chain management, and macroagriculture sector intelligence. The vast majority of these interventions are concentrated in the provision of advisory services and market access.

However, the sad reality is that pre-COVID-19, the number of registered versus active users of these digital innovations was limited. Many smallholder farmers and entrepreneurs in the ecosystem were unwilling or unable to pay for them. In addition, many of the companies that provide these solutions were struggling to scale even though they were benefiting from funding and support from development partners and financial institutions[4].

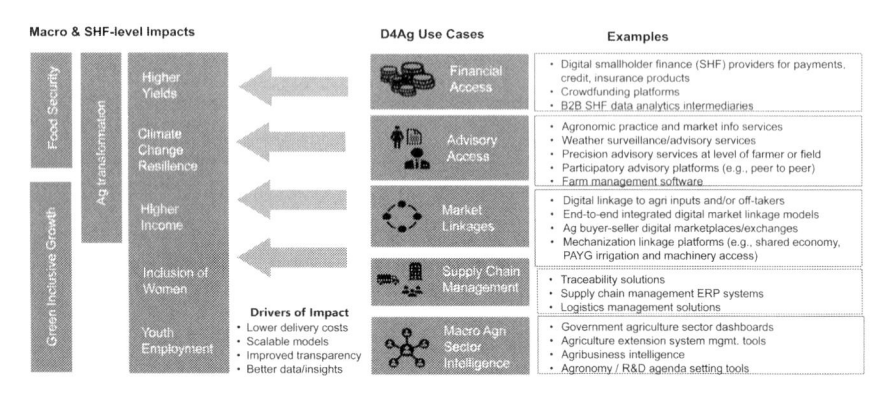

Figure 4.1 Spectrum of digital interventions on the African continent. Source: CTA. 2019. The Digitalisation of African Agriculture Report, 2018-2019. Accessed: https://www.cta.int/en/digitalisation-agriculture-africa (25 July 2020).

Benefits of embracing innovation and technology

What COVID-19 and other shocks have revealed is the critical need for entrepreneurs to understand and embrace the benefits of the innovations and technology solutions in the agriculture and food ecosystem. There are at least three significant benefits outlined below.

- **Increase your visibility:** The penetration of mobile devices and cellular and wireless networks is growing across Africa, enabling entrepreneurs to engage customers virtually. You can leverage technology to market your services or products through basic SMS, a strong online presence, and social media platforms. You can enhance your visibility via digital marketing through sponsored ads on websites, active engagement on social media platforms, or via regular e-newsletters to your mailing list. (This will be explored in greater detail in Chapter 5.)
- **Improve your service delivery:** Regardless of your business model, you can leverage technology to offer innovative or more convenient service options for your customers. You can improve your service delivery by leveraging data either through internal or external sources to identify and monitor related trends and their effects on your services. This will enable you to better tailor your services to fit the needs of your target customers.
- **Reduce your operating costs:** You can leverage various technological advancements to increase the productivity of your team members and operations and reduce your costs. First, you can automate business processes that might form a bulk of your running costs or transactions that are repetitive, such as reoccurring orders and payments. Embracing the use of mobile applications and online software could help save time and ensure that employees can channel their energy into other tasks and eliminate the risk of human error that may come at a cost to your operations.

You can even consider operating an online platform in the initial stages of your business to eliminate rental costs for an office space. In addition, utilizing applications such as Skype, FaceTime, Zoom, or WhatsApp for engaging with clients who are in distant locations can save travel costs.

WORKSHEET 4.1: ASSESSING YOUR COMPANY'S USE OF AGTECH, DIGITAL TOOLS, AND INNOVATION

How do you currently infuse digital tools, innovation, and technology into your operations?
- What digital tools, innovation, and technology do you use to reduce your costs?
- What digital tools, innovation, and technology do you use to increase your visibility?
- What digital tools, innovation, and technology do you use to enhance your service delivery?

Are you interacting with your suppliers online or via SMS?
- Are you tracking your input purchases and applications, leveraging some sort of digital platform?

Are you interacting with your existing and potential customers online or via SMS?
- Are you tracking your clients'/customers' purchases so that you can proactively reach out to them when they will need more products or services?

For entrepreneurs in primary production:
- Are you taking pictures of pests on your farms to get guidance from experts via phone?
- Are you interacting with input suppliers and markets online or via SMS?
- Are you using low-cost sensors to track on-field data (e.g., soil moisture, temperature, light indices, nutrient uptake)?
- Are you leveraging agtech for harvesting, postharvest management and storage?

Who manages innovation and technology within your organization?

Which organizations do you partner with to enhance your capacity in this area?

How are you building a culture of innovation and technology within your company?

Managing the costs and risks of utilizing digital tools and agtech: Entrepreneurs are often concerned about the cost implications of investing in technology, innovation, and digital tools. In reality, there are many free tools, cost-sharing opportunities, and support provided by development partners, government agencies, financial institutions, and others in the ecosystem.

Regardless of which approach is available to your company, guard against data and privacy issues and restrictions on your ability to partner with other players.

Technology, innovation, and digital tools through the value chain

There are a range of entrepreneurs leveraging innovation and technology to deliver products and services that support stakeholders across the value chain and enhance the productivity, efficiency, and effectiveness of the sector and enable scaling. Review the innovations and technology interventions outlined below to determine which to apply to your business model as you scale.

Table 4.1 Examples of Players in the Agtech/Digital Ag Ecosystem

Advisory and Information Services	Market Linkage Solutions	Digital Supply Chain Management services	Macro Agricultural Intelligence	Financial Access
Agdevco	CowTribe	CropIn	AWhere	Babban Gona
AgriGrow	Farm Crowdy	Farmz2u	Dataverse	Bokomaso
Agripadi Farm Services	FarmIT	iProcure	Agro data	Investment Impact
Agro Info Tech	Good Nature Agro Zambia	Kitovu	Crop Watch	Cellulant
BIC farms Concepts	Hello Tractor	Kobo360	Gro Intelligence	Efarms
Biakudia Urban Farming Solutions	Kobiri	Logistimo	Satelligence	One Acre Fund
Green Sahara	M-farm	Lori		Seekewa
iCow	mAgri	Metajua		Shambapro Limited
iShamba	myAgro	Virtual City		Thrive Agric
Justagric Agros	Traxi	Wakati		
Plug 'n' Grow	TroTro Tractor			
Rico Gado Agritech	Tulaa			
Rural farmers' hub				
Ujuzi Farmer				
WeFarm				

Agricultural inputs

Over the past few decades, advancement in input technology, mostly new and improved seed varieties and fertilizers, has led to increased crop productivity across the world. However, on average, smallholder farmers in Africa still obtain less than 20%–30% of the yields of their counterparts in other world regions. This gap in productivity presents a massive opportunity to infuse technology in food production and transform the agriculture sector in Africa. Similar to the way that mobile phones spread rapidly through the continent, allowing them to leapfrog landlines almost entirely, applying technology and innovation to agriculture and food systems will enable stakeholders in the sector to effectively increase food production. This will also strengthen food systems and boost profitability while overcoming many of the challenges that they face, ranging from limited access to quality inputs and high costs of production to poor marketing and last-mile distribution channels.

Seed technologies: Agricultural input providers such as seed companies, in collaboration with research institutions, are leveraging technology to develop new and improved seed varieties with characteristics including, added nutritional content; the ability to withstand effects of climate change, such as harsh weather conditions including drought and flood; and tolerance to diseases and pest attacks.

These innovations present opportunities, not just for farmers to leverage these seeds to enhance their productivity and profitability, but also for aspiring entrepreneurs to become foundation or certified seed producers or agrodealers, ensuring that high-quality seeds are available and providing last-mile distribution.

Research institutions such as the International Institute of Tropical Agriculture, working through a consortium of partners, has enabled the emergence of a growing Early Generation Seed (EGS) System for cassava and yam in Nigeria through a range of initiatives between 2015 and 2023. In collaboration with local research institutions in Ghana, Nigeria, and Tanzania, and African entrepreneurs, their interventions have revealed the untapped opportunities for the commercialization of indigenous research and have created opportunities for entrepreneurs across the ecosystem.

Box 4.1: CASE STUDY – EARLY GENERATION SEED (EGS) SYSTEMS FOR THE CASSAVA VALUE CHAIN[5]

Building a Sustainable, Integrated Cassava Seed System in Nigeria (BASICS) was implemented by the CGIAR program on Roots, Tubers and Bananas, led by the International Potato Centre, in partnership with the International Institute of Tropical Agriculture (IITA), National Agricultural Seeds Council (NASC), National Root Crops Research Institute (NRCRI), Catholic Relief Services (CRS), Context Global Development (CGD), and Fera Science Limited (Fera) between 2016 and 2020.

Before BASICS, there was no formal seed sector for cassava stems. Farmers would usually share cassava stems with their fellow farmers. In some cases, governments would buy improved stems and share them with farmers free of charge. However, most of the stems shared by the government were usually not certified and in small quantity because of the bulky nature of cassava. Such distributions happened only occasionally, either to address some exigencies or to introduce new varieties, and these too were targeted at certain segments/locations of farmers. This approach proved unsustainable as the spread of improved varieties still hovered at about 40%. The project director, Dr. Hemant Nitturkar, explained that initiative linked breeders and researchers who developed improved cassava varieties and technologies with farmers and processors who benefited from high-quality planting materials. According to Nitturkar,

> The BASICS project has created over 150 community-based seed entrepreneurs, called Village Seed Entrepreneurs (VSEs) who are running viable cassava stem businesses selling certified cassava stems to the farmers in states such as Benue, Cross River, Abia, Akwa-Ibom and Imo in Nigeria; and facilitated the establishment of two commercial seed companies – IITA GoSeed ltd located on the IITA campus in Ibadan, and Umudike Seed at National Root Crops Research Institute (NRCRI) in Umudike, Abia State – to ensure a reliable supply of breeder and foundation seeds of varieties in demand to these VSEs.

BASICS encouraged the development of VSEs, to locate seed production closer to the cassava growing communities and reduced the high costs associated with transporting bulky roots. These village seed entrepreneurs multiplied improved stems, and they sold certified seeds to the farmers on a commercial basis. The BASICS project, based on the feat accomplished in Phase 1 was given the opportunity to continue with a Phase 2 approved by the donor to concretize the outcomes of the project.

Another key innovation in seed systems is known as bioengineering, which applies the principles of biological and physical sciences to manipulate or impact the genetic traits of a crop to improve its performance and nutritional content. Biofortification helps increase the amount of dietary vitamins and minerals in staple crops, enhancing their nutritional value and optimizing production levels. Studies by HarvestPlus[6], the leading global nonprofit organization engaged in pioneering biofortification, show that conventional crop bleeding can increase nutrient levels without compromising yields.

Box 4.2: BIOFORTIFICATION – HARVESTPLUS' PIONEERING WORK IN SEED SYSTEMS

HarvestPlus works across Africa to catalyze the adoption, production and consumption of vitamin A cassava, iron bean, provitamin A (PVA) maize, and orange-fleshed sweet potato (OFSP). HarvestPlus focuses its biofortification efforts in several African countries including Nigeria, Rwanda, the Democratic Republic of Congo, Uganda, and Zambia. In Nigeria, the organization works with the public and private sector and international partners, such as IITA, to multiply vitamin A cassava stems and distribute to farmers for planting. It is also creating and strengthening demand by supporting commercial processing of vitamin A cassava into local food products such as garri and fufu. It has launched several awareness campaigns, leveraging mass media and curating a short film with the Nigerian entertainment industry, to educate Nigerians on macronutrient efficiencies and the benefits of vitamin A cassava.

From 2004 to 2020, HarvestPlus enabled the release of 242 varieties of 11 biofortified crops in 30 countries. The 11 biofortfied crops (iron beans and pearl millet; vitamin A cassava, maize, and sweet potato; zinc maize, rice, and wheat) are extremely important to the dietary

> requirements for women and children within the populations where the crop is consumed. For countries that are heavily reliant on these crops, biofortified iron beans and iron pearl millet can provide up to 80% of the daily average requirements. Zinc wheat and zinc rice can provide up to 50% of the average daily zinc needs, vitamin A cassava and sweet potato can provide up to 100% of average daily vitamin A needs. These biofortified crops show that it is possible to increase micronutrient levels without compromising the production of other crops.

Genetically modified organism (GMO) seeds have received mixed receptions on the African continent. While they are recognized for their resistance to pests, drought-tolerance, and higher yields, there are significant concerns around their costs and medium- and long-term health risks, given the uncertainties around the impact of GMOs. Three African countries have fully embraced GMO seeds – Burkina Faso, South Africa, and Sudan. According to the *African Business Magazine*[7], South Africa started with GMO maize in 1996, then cotton in 1997, and soybean in 2001. The entire market in Africa was valued at around $615.4 million in 2018, and is forecasted to grow by 5% annually to reach an estimated $871m by 2025. It is important to recognize that many countries in Africa have actively rejected GMOs. Thought-leaders and policymakers are concerned about Africa becoming a dumping ground for GMO seeds that are not allowed in other parts of the world, the long-term implications on the land, the environment, and health impact on farmers and the populace.

Soil health and fertility: Given the importance of soil health and fertility to ensure high yields, many entrepreneurs and farmers have embraced opportunities in this landscape by producing and retailing products and offering a range of support services. At one end of the spectrum, entrepreneurs managing soil testing laboratories are helping farmers determine what seeds and fertilizer would be most suitable for specific soil types and, at the other end, blending plant managers that produce a range of fertilizer offerings are partnering with agrodealers to ensure last-mile distribution.

There is also considerable innovation in homegrown solutions. For example, Ecodudu[8] in Kenya uses an innovative model that collects food waste and uses the black soldier fly larvae to recycle it into organic fertilizer and processes the larvae into animal feed.

> **Box 4.3: OCP'S PIONEERING EFFORTS IN THE AFRICAN FERTILIZER LANDSCAPE**
>
> Established in Morocco, with over a 100-year history, OCP is a leading global provider of phosphate and its derivatives. It develops fertilizer solutions customized for local conditions and crop needs. OCP Africa works with governments, nonprofits, and private companies in 18 countries. It specializes in creating customized products that require a deep understanding of the soil-crop-environment system and the farmer's practices. The approach is based on three levels of information. Firstly, OCP conducts an in-depth assessment of soil and crop responses. Through soil testing, soil analysis, soil fertility data, and onsite field trials, testing the crop response to developed fertilizers, OCP is able to customize their solutions according to the needs of the targeted region. The company also partners with local research and agronomy institutes in an effort to assess and produce the crops. Secondly, OCP engages in assessments of the current agronomic practices, which entails the collection of macroinformation on the farmers and their environment. The third stage of information is geospatial technologies for developing nutrient management platforms.
>
> OCP School Lab provides demonstrations for farmers with technology from its partner, SoilCares. These "labs," typically placed in mobile containers, travel to meet farmers where they are, work with them to test their soil, and provide fertilizer application recommendations for their soil and crop mix. As of December 2019, the School Lab had assisted more than 350,000 farmers in Kenya, Nigeria, Togo, Burkina Faso, Ghana, Tanzania, and Côte d'Ivoire.
>
> Agribooster, which was first launched as a pilot in Côte d'Ivoire in 2017, connects farmers to financing and insurance, works with local extension agents to train them on proper fertilizer use, and collaborates with other providers to ensure improved farmer productivity.

Crop protection: With the widespread incidence of crop pests and disease across Africa, farmers increasingly recognize the importance of crop protection. Historically, multinational companies such as Syngenta, UPL, Monsanto, BASF, Adama, Sumitomo, and Corteva have dominated the provision of crop protection products on the continent. However, in recent years, a few African entrepreneurs, such as the Candel Company[9], established by

Charles Anudu, have entered the formal sector. Candel produces a range of herbicides, insecticides, fungicides, and growth stimulants.

Widespread uptake among smallholder farmers remains relatively low, and farmers are often wary of the genuineness of products, given the extensive sale of adulterated products in the market, which are usually less effective.

There are also significant concerns about the health risks associated with many of the crop protection products currently available in African markets and the limited farmer education on effective methods of usage and storage. A study by researchers at Cornell University and Georgetown University[10] of farmers in Ethiopia, Nigeria, Tanzania, and Uganda revealed evidence of a range of negative human health outcomes associated with pesticide use, including time lost due to illness.

Sadly, many farmers do not have protective clothing and equipment to minimize contact. However, there are emerging technologies beyond the use of backpack pesticide sprayers, which include deploying drones that ensure exact doses and targeted applications of products, with less direct exposure. Drones are being utilized on a few commercial farms on the continent, and ultimately, there is significant room for innovations in products and application methods.

Delivery of inputs: In order to improve access to inputs, online marketplaces are emerging to enable growers to access quality inputs that will boost their productivity and increase profitability. Afrimash[11] in Nigeria is one example of an online marketplace that provides access to livestock inputs, such as poultry and fish, poultry equipment, and veterinary services.

Mechanization: Technological advancements in farm machinery ensure efficiency of labor in farm operations, from land preparation to crop cultivation, irrigation, harvesting, and processing.

In 2014, Hello Tractor[12] was established to improve the access of smallholder farmers to timely and affordable tractor services. The company began its operations with the direct sale of tractors to farmers and cooperatives, who then utilized its mobile sharing platform to identify and meet the demand for the use of tractors by other farmers. However, in 2017, Hello Tractor decided to leverage technology and focus more on the development of its mobile platform. Through its platform, the company enables farmers to request affordable equipment while providing security to tractor owners through remote asset tracking and virtual monitoring, earning it the name "Uber for Tractors."

Currently, Hello Tractor is using data to enhance its service delivery and is collaborating with IBM to pilot an advanced agricultural analytics and decision-making tool that cuts across the mechanization ecosystem – inclusive of farmers, tractor owners, financial institutions, and governments. Through artificial intelligence and the IBM blockchain, farmers on the Hello Tractor platform can obtain relevant information to increase their yields. Tractor owners can receive insights to save time and earn more, and banks have access to information to help them determine a credit portfolio for both the farmer and the tractor owner.

Production

There are a range of innovations available for entrepreneurs engaged in crop production.

- **Sensors:** Enables the application of big data and analytics in agriculture. By providing farmers with real-time information about the state of their crop, livestock, soil, or farm machine, sensor technologies ensure more significant levels of consistency and reliability in decision making to optimize production.
- **Automated technologies:** Includes drones, unmanned aerial vehicles (UAVs), robots, and artificial intelligence (AI) used to perform agricultural processes accurately and with limited involvement of humans.
- **Vertical farming:** The practice of growing produce in vertically stacked layers. The practice can use soil and water or be soilless, leveraging hydroponic or aeroponic growing methods.
- **Blockchain technology:** Enables the traceability of information and data in the food supply chain and helps improve food security.
- **High-roofed greenhouses:** A great way to increase production; however, they require significant amounts of land. Studies show that greenhouses with roofs 12-feet or higher have doubled farmers' yields.

Leveraging precision agriculture, farmers receive tailored information on the use and application of inputs, appropriate planting times, and most fertile areas of their land for crop types to improve their productivity, resource efficiency, and minimize costs. Zenvus[13], a Nigerian company, uses precision monitoring tools and data analytics tools to analyze soil data such as temperature and nutrients. This ensures that farmers are informed on the appropriate fertilizers to apply and time to irrigate their farms, reducing input waste and improving farm productivity.

Beat Drone[14], a Nigerian startup, utilizes drone and online application technology to support various farming activities. The firm uses drones as a spraying tool on farms to engage in crop supervision, map farmlands to improve agricultural yields, and reduce labor dependencies. On the company's platform, a farmer can request a drone, schedule a date, and make payments before the drones are deployed.

After planting, farmers can continue to leverage technology to grow their crops effectively and aid the timely identification of diseases. For example, Saillog[15] leverages artificial intelligence through its mobile app that enables farmers to identify and treat plant diseases. Farmers can upload a picture of their afflicted plants to the platform from their smartphone, and experts from around the world can help diagnose the diseases and offer treatment solutions. For farmers without a smartphone, WeFarm[16] in Kenya provides an alternative option through its farmer-to-farmer digital network that enables farmers to send questions to and receive answers from other farmers in their global network, all via SMS.

Extension support/service provision for farmers: There are a range of entrepreneurs in the agriculture landscape that provide bundled support services to farmers through the production phases. For example, Kola Masha established Babban Gona in 2012 to provide enhanced agricultural technologies and end-to-end services that optimize yields and labor productivity, while simultaneously improving market access for smallholder farmers[17]. The Babban Gona team provides farmers with financial services, agricultural and input services, training and development, and marketing services. The combination of these offered services enable farmers to triple their yields and enhance their livelihoods.

AgroSpaces[18], founded by Jim Bakoume and based in Cameroon, provides a marketplace system that connects farmers, buyers, consumers, and other agroactors to share information and form valuable connections. The platform provides farmers with access to agriculture news, announcements, market information, agricultural tips, and region-specific weather forecasts. With a database of more than 12,000 young startups, AgroSpaces is changing the mindsets of youth and drawing them into the agriculture ecosystem.

Cowtribe[19], an agritech company based in Ghana, delivers life-saving veterinary services to farmers that allows them to manage livestock vaccination cycles and ensure onsite delivery. Since its inception in May 2016, Cowtribe has grown its customer base to more than 31,000 farmers, with more than 1,000 of them registered as subscribers. Cowtribe's business model allows farmers to spend less of their income on vaccines for their livestock, thus allowing farmers to invest more in their families and build their savings.

Arifu[20], another social enterprise headquartered in Kenya, has an innovative chatbot platform for engaging, training, and capturing insights for important and hard-to-reach audiences, including rural farmers. The learning content is accessible on both basic and smartphones, with or without internet, over SMS and chat apps. Arifu partners with development organizations and private organizations to provide access to the relevant capacity building and information on products and services so that no SMS fees or content subscriptions are imposed on the user.

Big data: The application of big data in agriculture has the potential to provide farmers with comprehensive data on rainfall patterns, impending drought, and fertilizer requirements and enables them to optimize their farm equipment and manage their supply chains. This empowers the farmers to make smart decisions, such as what crops to plant, how best to maintain machinery, and when to harvest, all aimed at improving yields, reducing postharvest losses, and maximizing profitability.

The CGIAR – Platform for Big Data in Agriculture aims to "harness the capabilities of big data to accelerate and enhance the impact of international agricultural research."[21] The platform, which is part of a 5-year project, provides global leadership in organizing open data, convening partners to develop innovative ideas, and demonstrating the power of big data analytics through inspiring projects. Through this platform, CGIAR aims to increase the impact of agricultural development through adapting knowledge from big data to solve development problems faster, better, and at greater scale than previously done.

Gro Intelligence, operating from both the United States and Kenya, brings the power of forecasting to agriculture-related challenges. By organizing global data, the platform allows stakeholders, including input companies, the financial sector, food and beverage companies, agribusinesses, and others to quickly find data as well as integrate their data, build predictive models, derive meaningful insights, and make better informed agricultural decisions.

Financing farmers using digital tools: There is an emerging ecosystem of entrepreneurs providing financing for farmers leveraging digital tools.

In 2001, Ken Njoroge and Bolaji Akinboro cofounded Cellulant[22] with a personal investment of $3,000. From offering data and e-wallet systems for farmers to access seeds and fertilizer, Cellulant has evolved into a digital payments company. One of its core products, Agrikore, is a blockchain-based smart-contracting, payments, and marketplace system that connects key stakeholders in a trusted environment. These stakeholders include farmers, fast-moving consumer goods companies, inputs providers, produce aggregators, insurance companies, financial institutions, governments, and development partners.

Farmcrowdy Ltd,[23] founded by Onyeka Akumah, is Nigeria's first digital agriculture platform. Through Farmcrowdy, individuals can commit to an agreed upon sum to own a minimum farm space and start and complete a farming cycle. Farm partners on the platform can sponsor any farm of their choice, including maize, poultry, and cassava farms. They get biweekly updates about their farm's progress, including pictures and videos from the farmers. By 2019, Farmcrowdy had worked with more than 25,800 farmers, and recently launched an e-commerce food store that allows people to order fresh farm produce from the Farmcrowdy's network of farmers.

Mobbinsurance[24] is a fintech startup company based in South Africa that is focused on offering farmers affordable crop insurance using mobile applications. Established by Kudzai Kutukwa, Mobbinsurance also plans to utilize satellite imagery and weather data to protect farmers against weather-related risks that cause crop failure

Agriculture and Climate Risk Enterprise Ltd.[25] (ACRE Africa) links farmers to local insurers and other stakeholders in the agricultural insurance value chain. Operational in Kenya, Rwanda, and Tanzania, the ACRE Africa team, engage in risk assessment, product development, and risk monitoring to facilitate access to crop and livestock insurance products for smallholders.

Postharvest storage/aggregation/logistics

Researchers estimate that between 20% and 60% of produce, depending on the value chain, cultivated by smallholder farmers in Africa goes to waste[26].

This reality is linked to the archaic storage methods, the poor state of the road networks, limited cold storage infrastructure, and energy poverty. In fact, according to Joshua Sandler,[27] the founder of Lori Systems, "The relative cost of moving goods in Africa is one of the highest in the world, leading to up to 75% of a product cost's going to logistics (compared to 6% in the US)."

A range of initiatives and entrepreneurial ventures are working to address these critical challenges, which are vital for Africa to feed itself and the world. For example, the Global Alliance for Improved Nutrition (GAIN) created the Postharvest Loss Alliance for Nutrition (PLAN) to foster appropriate solutions to reducing high postharvest losses caused by lack of functional cold chain systems and inappropriate packaging of perishable produce. This membership organization has spurred knowledge sharing and innovations in packaging and cold storage. One of its members, ColdHubs[28] offers farmers reusable crates, which fit into the storage devices developed by the company. Leveraging a flexible pay-as-you-store subscription model, farmers pay a daily flat fee for each crate of food they store in a solar-powered walk-in cold room, for 24/7, off-grid storage and preservation of perishable foods.

InspiraFarms,[29] founded in 2012 by Zimbabwean Tim Chambers and Italian agronomist Dr. Michele Bruni, provides small and growing agribusinesses with the tools, technology, and expertise to significantly reduce food loss and energy costs. The company is a first-mile cooling technology distribution and financing company that supports businesses in East and Southern Africa.

Beyond storage, innovative solutions are emerging across the continent focused on aggregation and logistics and leveraging technology. In the Nigerian context, there is AFEX,[30] whose mission is to create the right incentives and a shared platform to change the food systems in Africa – making them more inclusive, scalable, and efficient, with shared prosperity and trust embedded. According to AFEX CEO, Ayodeji Balogun – "AFEX has since reached and enhanced the livelihoods of over 100,000 farmers and aggregated 100,000 metric tons of grains with the organization's overarching strategy of a national trading platform and supply chain network in carefully identified value chains."[31]

The Ethiopian Commodity Exchange (ECX)[32], established in 2008, is a marketplace, whose mission is to connect all buyers and sellers in an efficient, reliable, and transparent market by harnessing innovation and technology and based on continuous learning, fairness, and commitment to excellence. Leveraging warehouse receipt systems, digital platforms, and a formal exchange connected to international markets, ECX trades coffee, sesame, white pea bean, green mung beans, soya bean, and sesame.

> **Box 4.4: LOGISTICS IN AFRICA – KOBO360**
>
> Kobo360,[33] established in December 2017, is a technology company that aggregates end-to-end haulage operations to help cargo owners, truck owners, drivers, and cargo recipients to achieve an efficient supply chain framework. From a sole focus on the transportation of agricultural inputs such as fertilizer and grains for large, fast-moving consumer goods (FMCGs) companies, including Flour Mills of Nigeria and Olam, Kobo360 has grown dramatically in just 3 years.
>
> Kobo360 uses big data and technology to reduce logistics friction while empowering rural farmers to earn more by reducing farm waste and helping manufacturers of all sizes to find new markets. In 2020, it managed 17,000 trucks and operated in Nigeria, Ghana, and Kenya.
>
> In addition to its logistics services, Kobo360 offers, KoboCare a solution for registered drivers on the Kobo360 platform, through which it offers discounted diesel sales to Kobo360 drivers in partnership with Oando/Total filling stations, access to Diver Support Centers across Nigeria with services such as filling stations' maintenance bay, fire services, law enforcement dedicated parking, rest bay for drivers etc., and an HMO plan for drivers and their families. The company also provides KoboSafe, a self-insurance solution that provides insurance coverage for every single trip on Kobo360's platform.
>
> In partnership with companies such Cellulant, ThriveAgric, Babban Gona, Kobo360 is exploring logistical support for aggregation from smaller farmers. The company is also investing in the provision of cold chain logistics, warehousing, and infrastructure with support from a range of local and international partners and the International Finance Corporation (IFC).

Processing

Processing is arguably one of the most critical stages of the value chain in Africa that can contribute to agricultural transformation. Adapting technology to increase the efficiency of food processing on the continent can reduce postharvest losses, ensure value addition and year-round availability of food, lower the cost of nutritious food, increase food safety, and spur significant economic growth.

Given the high level of microbial load generated during outdoor drying, a method typically utilized by many smallholder farmers, commercial steam sterilization is an effective method used by companies such as AACE Foods in Nigeria, to reduce pathogens in food products.

In terms of packaging solutions, there are a few emerging innovations being spurred by companies in South Africa, such as Green Home Packaging[34] that sells 100% biodegradable food packaging made from plants. The company's mission is to change some trends, rewrite some rules, and make plant-based food packaging the norm by 2030. Other companies innovating packaging solutions include Lovell Industrial,[35] which specializes in a variety of thermoformed and die-cut packaging designed to enhance food's visual appeal, and Nature Pack,[36] an ecofriendly packaging, manufacturer and supplier company.

Market linkages

Access to markets remains one of the biggest challenges that farmers face across Africa. Thankfully, there are many emerging success stories of entrepreneurs leveraging innovation and technology to ensure that farmers can access markets. For example, TruTrade Africa,[37] based in Uganda, operates as a social enterprise, providing smallholder farmers with a reliable route to market and fair prices for their produce. The company does this through its "Trade Transparency Solutions," aimed at making rural agricultural markets function more efficiently for farmers, aggregators, and buyers. As part of their farm to fork services, TruTrade negotiates supply contracts with buyers through their online platform. Thereafter, the team sources from farmers who bring their produce to one of their multiple collection points where agents can assess the quality, weight, and instantly pay through the TruTrade app to the farmers' mobile or bank account. TruTrade also manages the aggregation from different agents, transaction logistics, and delivery to the final buyer. The company currently engages in oilseeds, cassava, tree crops, cereals, mung beans, and poultry.

Farm Fresh,[38] which was launched in 2014 as the first online fruits and vegetable store in the Gambia, connects a network of over 20 fruits, vegetables, and livestock farmers across the country, particularly in the rural areas with customers in the cities. Founder and CEO Modou N'jie identified a gap between farmers and consumers and decided to leverage his two decades of experience in IT management to create linkages.

Globally, agrifood tech companies focusing on retail are receiving more funding from investors than any other segment ($10.1 billion vs. $3.5 billion for agtech companies focused on both input providers and growers).[39] Retailers are using technology to provide customers with personalized experiences and products as well as to deliver products to customers more efficiently and on-demand.

Virtual Farmers Market (VFM) is an app-based e-commerce platform in which farmers can trade and sell the surplus. VFM was a program developed by the United Nation's World Food Program in July 2016.[40] The app was designed with the Airbnb and eBay digital models in mind – to provide a platform to match smallholder farmers and consumers.

A pilot version of the app called Maano was launched in Zambia in 2017. By connecting producers and consumers, Maano has been able to create a more direct connection between buyers and sellers. This reduces transaction costs for both parties and removes barriers to market entry for small-scale farmers that would not typically have the financial or human resource capacity to create and run their own e-commerce platforms. In 2017, the app reached more than 1,000 farmers in Zambia, with a total of 100 transactions.[41]

Surviving and thriving as an agtech business service provider

If you are an entrepreneur whose business model is firmly rooted in leveraging digital tools, innovation, and technology to solve problems in the ecosystem, your role in this ecosystem is critical to enable effective leapfrogging. As you scale your operations, there are challenges that you must overcome.

- **Developing a compelling product/service offering and a business case:** Barriers to entry in this sector appear low, and there are many me-too solutions, effectively copying early movers. As a result, you have to commit to:
    - **Before launch** – Focus on developing products through active engagement with target customers, ensuring the design of products/services that are affordable, accessible, and acceptable. This

will ensure that the final product or service that you design meets the needs of its intended customers or clients, is user-friendly, and does not require extensive or unnecessary behavior change. Iterating on design and interviewing customers and clients is critical. In addition, registering the intellectual property (IP) of the design or the approach and actively developing clear strategies for protecting this is critical.

It is also imperative to determine a clear pricing strategy, which leverages a range of options, including pay for service, pay in advance, and payment by a third party. For example, weather, market-linkage, and knowledge-transfer services that serve smallholder farmers are often paid by the off-takers who value the traceability, predictability, and data that can be obtained from the farmers.

- **After launch** – You must regularly monitor the activities of current and potential competitors, usage, and market access. Competitor offerings, depending on what is protected by IP, can be feasible. Finally, identify what steps should be taken to address these barriers. Can government support or aggregating demand help? Can you refine the product with more input from customers? Following these steps will help to increase stickiness once a product or service is already in the market.
- **Attracting, retaining, and scaling paying customers/subscribers:** As reinforced by CTA's research and other industry studies, startups in the African agtech ecosystem struggle to attract and retain paying customers. For those who are successfully scaling, they are utilizing a range of approaches, including the following:
  - **Free trials:** This attracts potential customers, and if they value the services that they are receiving, it will induce them to switch to paying customers after the trial period is over.
    - For services that require a subscription, a limited-time trial is preferred. Another option is offering a "freemium" version, in which a basic skeleton of the service is free for customers, and additional features are available for purchase (especially standard for software apps).
      - Using data and technology, such as QuickTrials,[42] which is an online software that allows users to create online

and paper surveys and software forms and then collates, aggregates, analyzes, and generates reports and findings. This reduces the amount of time spent on collating data and formulating reports. This robust software with multiple features offers free trials lasting 14 days upon subscribing. (https://www.quicktrials.com/)
- For products that are consumable (most consumer-packaged goods and/or seasonal inputs for growers), a small sample can be useful in increasing demand, especially for newer products. Selecting the right amount for the sample is critical because if the amount is too small to be effective, it will not increase demand, but if it is too large, customers may no longer need to buy the product.

  For products that are not consumable (large machinery, etc.), offering a limited-time or a group trial can be effective in increasing demand.
- When considering agrifood-related enterprises, it is important to note that the purchasing, testing, and validation cycles depend on the length and frequency of the growing season. For on-farm products and services, offering a sample in a portion of a grower's land could be a compelling way to increase demand.
- The model customer user approach is similar to the free trials suggested above, for conducting a pilot, but relies more on network effects from customers' peers to increase demand. It typically involves creating pods of five to six users, with one designated as the "model user" that is responsible for training his or her peers on the value of the product/offering. In selecting each model user, it is important to identify individuals who are more willing to incorporate new and innovative products into their traditional processes and to identify individuals who wield a certain amount of formal or informal influence in their preexisting social groups. This is a common method used by health and agriculture extension workers in Ethiopia and can be made even more effective by ensuring that the model user is one with similar characteristics to target customers, including reliance on different sources of information and local experts or leaders.

- Free trainings: For products or services that may require behavior change from customers, providing free trainings is paramount to increasing demand. This is often utilized for primary producers through farmer field days and for software users through invitation-only sessions for potential customers. Repeated training and customer support will help encourage the behavior change from customers, and upon seeing time saved or improved results, these customers will recruit their peers to the product or service as well.
- **Measuring impact/value creation:** There are a number of ways to determine whether your innovation is achieving impact.
    - First, engage in a baseline study via focus group testing or customer surveys. Also, include a control group. This should be iterative and a major part of the product development life cycle and is critical to conduct prior to launching to ensure that the product/offering is user-friendly and addresses its intended users' pain-points. Before a large-scale investment and launch, piloting the product/offering is also a suitable method of assessing initial demand.
    - Second, track how many users are voluntarily marketing it to their peers via word-of-mouth. If an innovation's customer acquisition cost exceeds its customer lifetime value, then the innovation is not likely to survive because it either does not meet the customers' needs effectively and/or it is inaccessible to them because of its cost.
    - Third, offer a free trial to users and see if customers are willing to pay for it after the trial has ended. This indicates that the value they derive from it is also greater than its cost.
    - Fourth, actively capture data on customer usage, behavior, habits, and needs, and assess it on a weekly basis. This data could prove valuable in enabling you to effectively meet the needs of your customers/clients and enhancing other aspects of your operations. It will also improve your ability to engage with customers either within or outside the industry. It could also be a valuable product to share with others, contingent on the data privacy rules that you have agreed upon with your customers/clients. For example, Farmcrowdy's food platform captures data including users' search patterns, purchasing habits, and frequency, allowing the company to customize adverts and specials for each unique user as well as generate insights on products that are in high demand and those that may need to be reconsidered.

If and when you can demonstrate measurable impact, engage in strategic communications to amplify this impact. The tools for effectively raising awareness about your product/service and its effects will be addressed in greater detail in the next chapter.

You must make a concerted effort to engage female customers and clients, given the digital divide that exists in Africa, the cost of data and digital tools, and the barriers that women often face accessing technology tools and devices in a more traditional and male-dominated society. This also suggests the need to ensure that you involve female team members from the design and implementation phase, including the inclusion of female extension workers, especially when interventions require demonstrations aimed at female farmers.

Finally, you have to be extremely agile and adaptable to changes in your ecosystem, attacks, and shocks. Your ability to have robust scenario planning, invest in insurance, and actively preempt crises, including hacks on your software, where applicable, is critical to your long-term success.

Summary

As an entrepreneur, you can leapfrog, dramatically increase your productivity and profitability, scale your operations, and ensure that you can effectively achieve profitable growth hinged on your willingness to embrace digital tools, innovation, and technology. However, you have to commit to examining your business model to understand how you can leverage these tools to enhance your operations, increase your visibility, and maximize your profitability and impact. In addition, you have to actively monitor the immediate and long-term risks and costs associated with embracing some of these technologies. Finally, if your business model is rooted in the provision of agtech and digital tools, it is imperative that you develop a clear and compelling business case, and design appropriate and relevant products and services that are demand driven, sustainable, and resilient with measurable impact.

Notes

1   Forbes Online; High John, *Who Led Your Digital Transformation? Your CIO Or COVID-19?* May 26 2020; https://www.forbes.com/sites/peterhigh/2020/05/26/who-led-your-digital-transformation-your-cio-or-covid-19/?sh=32475305323a (Accessed: 12 November 2020).

2. Centre for Agricultural and Rural Cooperation. (2019) *The Digitalisation of African agriculture report, 2018–2019* [Online]. Available at: https://www.cta.int/en/digitalisation-agriculture-africa (Accessed: 25 July 2020).
3. Centre for Agricultural and Rural Cooperation. (2019) *The Digitalisation of African agriculture report, 2018–2019* [Online]. Available at: https://www.cta.int/en/digitalisation-agriculture-africa (Accessed: 25 July 2020).
4. Centre for Agricultural and Rural Cooperation. (2019) *The Digitalisation of African agriculture report, 2018–2019* [Online]. Available at: https://www.cta.int/en/digitalisation-agriculture-africa (Accessed: 25 July 2020).
5. "BASICS" box - Interviews with Dr. Hemant Nitturkar of BASICS and the team at IITA 2019–2020.
6. Harvest Plus. (n.d.) *Home page* [Online]. Available at: https://www.harvestplus.org/ (Accessed: 25 July 2020).
7. Kede, S. (2019) "GM foods: The battle for Africa," *African Business*, 20 November [Online]. Available at: https://africanbusinessmagazine.com/sectors/agriculture/gm-foods-the-battle-for-africa/ (Accessed: 1 August 2020).
8. Ecodudu. (n.d.) *Home page* [Online]. Available at: http://ecodudu.com/ (Accessed: 1 August 2020).
9. Candel. (n.d.) *Our mission* [Online]. Available at: https://www.candelcorp.com/index.php/mission/. (Accessed: 24 August 2020).
10. Sheahan, M., Barrett, C.B., and Goldvale, C. (2017) "Human health and pesticide use in sub-Saharan Africa," *Agricultural Economics, 48*(S1), 27–41. Available at: http://barrett.dyson.cornell.edu/files/papers/Sheahan%20Barrett%20Goldvale%20-%20SSA%20pesticide%20and%20human%20health%20paper%20Mar%202017%20final.pdf
11. Afrimash. (n.d.) *Home page* [Online]. Available at: https://www.afrimash.com/ (Accessed: 24 August 2020).
12. Hello Tractor. (n.d.) *Home page* [Online]. Available at: https://hellotractor.com/about-us/ (Accessed: 1 August 2020).
13. Zenvus. (n.d.) *Home page* [Online]. Available at: https://www.zenvus.com/ (Accessed: 1 August 2020).
14. Beat Drone. (n.d.) *Home page* [Online]. Available at: http://beatdrone.co/ (Accessed: 1 August 2020).
15. Saillog. (n.d.) *AgrioShield* [Online]. Available at: https://www.saillog.co/agrioShield.html (Accessed: 11 August 2020).
16. WeFarm. (n.d.) *Home page* [Online]. Available at: https://wefarm.co/ (Accessed: 11 August 2020).
17. Babban Gona. (n.d.) *Mission statement* [Online]. Available at: https://babbangona.com/mission-vision-and-values/ (Accessed: 1 August 2020).

18 AgroSpaces. (n.d.) *Home page* [Online]. Available at: https://agrospace.az/en/ (Accessed: 1 August 2020).
19 CowTribe. (n.d.) *Home page* [Online]. Available at: https://www.cowtribe.com/ (Accessed: 1 August 2020).
20 Arifu. (n.d.) *Home page* [Online]. Available at: https://www.arifu.com/ (Accessed: 1 August 2020).
21 CGIAR – Platform for Big Data in Agriculture Website; https://bigdata.cgiar.org/about-the-platform/ (Accessed: 11 November 2020).
22 Cellulant Nigeria. (n.d.) *Home page* [Online]. Available at: https://cellulant.com.ng/ (Accessed: 1 August 2020).
23 Farmcrowdy. (n.d.) *Home page* [Online]. Available at: https://www.farmcrowdy.com/ (Accessed: 11 August 2020).
24 Sahel Consulting Agriculture & Nutrition Ltd. (2018). *Agricultural technology* [Online]. Available at: https://sahelconsult.com/wp-content/uploads/2019/09/Sahel-Newsletter-Volume-19.pdf (Accessed: 24 August 2020).
25 ACRE Africa. (n.d.) *Home page* [Online]. Available at: https://acreafrica.com/ (Accessed: 1 August 2020).
26 RELOAD Project. (n.d.) *Post harvest losses – A challenge for food security* [Online]. Available at: http://reload-globe.net/cms/index.php/research/7-post-harvest-losses-a-challenge-for-food-security
27 Lori Systems. (n.d.) *Home page* [Online]. Available at: https://www.lorisystems.com/home (Accessed: 1 August 2020).
28 ColdHubs. (n.d.) *Home page* [Online]. Available at: http://www.coldhubs.com/ (Accessed: 1 August 2020).
29 InspiraFarms. (n.d.) *Home page* [Online]. Available at: https://www.inspirafarms.com/ (Accessed: 1 August 2020).
30 Associated Foreign Exchange Nigeria. (n.d.) *Home page* [Online]. Available at: https://afexnigeria.com/about (Accessed: 1 August 2020).
31 The Guardian Newspapers, Otaru Anthony, *AFEX to invest N2 trillion in agric sector growth,* 19 March 2020; https://guardian.ng/business-services/afex-to-invest-n2-trillion-in-agric-sector-growth/ (Accessed: 11 November 2020)
32 Ethiopian Commodity Exchange. (n.d.) *Home page* [Online]. Available at: http://www.ecx.com.et/ (Accessed: 1 August 2020).
33 Nsehe, M. (2019) "Q&A with Kobo360 co-founder Obi Ozor on his e-logistics startups's $30 million raise," *Forbes* 3 September [Online]. Available at: https://www.forbes.com/sites/mfonobongnsehe/2019/09/03/qa-with-kobo360-co-founder-obiora-ozor-on-his-e-logistics-startupss-30-million-raise/#968a82010e59. (Accessed: 11 August 2020).
34 Green Home Packaging. (n.d.) *Home page* [Online]. Available at: https://greenhome.co.za/ (Accessed: 1 August 2020).

35 Lovell Industries. (n.d.) *Home page* [Online]. Available at: http://www.lovell.co.za/ (Accessed: 1 August 2020).
36 Nature Pack. (n.d.) *Home page* [Online]. Available at: Accessed: http://www.naturepack.co.za/packaging-manufacturing.html (1 August 2020).
37 TruTrade Africa. (n.d.) *Home page* [Online]. Available at: http://www.trutradeafrica.net/ (Accessed: 1 August 2020).
38 Farm Fresh Gambia. (n.d.) *Home page* [Online]. Available at: https://www.farmfresh.gm/ (Accessed: 1 August 2020).
39 Crunchbase. (n.d.) *AgTech companies* [Online]. Available at: https://www.crunchbase.com/hub/agtech-companies. (Accessed: 24 August 2020).
40 Food and Agriculture Organization of the United Nations. (n.d.) *The International Symposium on Agricultural Innovation for Family Farmers: 20 success stories of agricultural innovation from the Innovation Fair*. Available at: http://www.fao.org/3/CA2588EN/ca2588en.pdf (Accessed: 1 August 2020).
41 Food and Agriculture Organization of the United Nations. (n.d.) *The International Symposium on Agricultural Innovation for Family Farmers: 20 success stories of agricultural innovation from the Innovation Fair* [Online]. Available at: http://www.fao.org/3/CA2588EN/ca2588en.pdf (Accessed: 1 August 2020).
42 QuickTrials. (n.d) *Home page* [Online]. Available at: https://www.quicktrials.com/ (Accessed: 1 August 2020).

5

BUILDING YOUR BRAND AND AMPLIFYING YOUR IMPACT

Introduction

Company names like Babban Gona, blueMoon, Java Foods, Farmcrowdy, Garden of Coffee, GBRI, Hello Tractor, Java Foods, One Acre Fund, Twiga, and Tula are repeatedly celebrated within Africa and across the globe. On face value, they seem to get more than their fair share of visits from international organizations, government support, investor and development partner funding, and customer and client engagement. Dig a little deeper. You will recognize that these organizations have strategically invested in building their brands and amplifying their impact in the local and international food and agriculture landscapes and even beyond the sector.

Sadly, only a few food and agriculture companies in the African context have gained the same continent-wide and international appeal. Instead, they struggle with telling their stories, demonstrating their impact, and increasing local, national, and international credibility. They often fail to counter the negative stereotypes and biases against them, which have historically painted Africa's food as substandard and our entrepreneurs as unethical. This limits the entrepreneur's ability to connect with clients and customers, attract local and international funding, and connect with a wide range of stakeholders based on trust and credibility.

This chapter will explore the strategies, tactics, and tools utilized by the leading companies. It will also examine steps for building your personal, company, and product/service brand to tell your story, inspire trust, and build credibility as well as attracting, exciting, and retaining customers, clients, investors, and partners. It will specifically address how to leverage media tools and strategic partnerships to build brands. Finally, it will examine how companies have effectively managed public relations crises and lessons that you can leverage.

Critical steps for developing your brand strategy

A brand is a promise kept, evident in the mind of consumers. Peter Njonjo, Cofounder/CEO Twiga Foods

A brand is a unique design, sign, symbol, words, or a combination of these, employed in creating an image that identifies a product or service and differentiates it from its competitors. Over time, this image becomes associated with a level of credibility, quality, and satisfaction in the consumer's mind.

There are many critical components of brand building and messaging that you must consider when starting and growing your company in the food and agriculture sector. Many of these issues are not unique to this sector, but a few are.

Your company name: Selecting a name for your company is often the first task for an entrepreneur. Many companies distinguish between their company name and the brand names of their products in the food and agriculture sector. In determining both, ensure that you select names that are memorable, purposeful, appropriate, and available.

- **Memorable:** The company name should be catchy and easy to remember and recognize. Hello Tractor was selected by the founder's wife, Martha Haile, because it would be easy for the average farmer to remember. Similarly, Mira Mehta chose the name Tomato Jos for two reasons; firstly, when she first moved to Nigeria in 2008, she was introduced to an inspiring group of entrepreneurs who encouraged her to develop and nurture what she describes as a "random pipe dream." The members of this group were proud Jos natives. Then when then the popular song "Oyi" by Nigerian artist Flavor was released, Mehta capitalized on the lyrics of the song, which had a play on words linking the best fruits and vegetables, in this case, tomatoes to Jos.

- **Purposeful:** Its meaning or symbolism should resonate with the stakeholders in your ecosystem. *Babban gona* means "Great farm" in Hausa, a language spoken predominantly in Northern Nigeria, and this name reinforces the mission of the company and its positioning within the ecosystem.
- **Appropriate:** Given the diversity of languages, religions, and cultural norms on the African continent, ensure that the names that you select are appropriate for most communities. Check their meanings in diverse languages. One word can mean "joy" in one language and "sorrow" in another.
- **Available:** Always check with the company registration agency in your country and the regions in which you plan to operate to ensure that another company has not utilized the name. Also, ensure that another organization has not trademarked the name.

Box 5.1: SELECTING A BRAND NAME FOR AACE FOODS

When Mezuo and I first tried to incorporate our food company, we chose AACE, an acronym for African Alliance for Capital Expansion (AACE), an investment club that we had established in our early 20s. We selected this name because it symbolized our hopes that the company would generate impact in the African context. It brought back memories of our initial foray into entrepreneurship. Not surprisingly, the Nigerian Corporate Affairs Commission would not accept a simple name. After multiple attempts and iterations, the agency eventually agreed to the name – AACE Food Processing & Distribution Ltd.

At this time, we also recognized that AACE would not be suitable as our company's brand name. We selected Ona, which means "jewel" or "precious" in the Igbo language, symbolizing our heritage as Igbos but was also simple and easy to remember. We immediately developed a compelling design and color for both the logo and brand name and advertising materials. We believed that this name was memorable,

> purposeful, and appropriate. We applied for trademarks for both AACE Foods and Ona.
>
> Our names and marks were published in the Trademark Journal, and soon after, we received a threatening letter from a respected Nigerian law firm that stated that we should release the Ona brand name because it was too close to their client's brand name *ONGA*. This letter was followed by a barrage of telephone calls from the parent company in South Africa. We were perplexed and destabilized, given our early investment in developing the brand materials. However, after engaging our board, they strongly advised that, as a startup, we could not afford to invest the funds and the time required to fight for a name that was not yet known. We decided to forgo *Ona* and proceed with AACE Foods as the brand name.
>
> Interestingly, the company that attempted to sue us ended up becoming one of our early institutional customers, and we were able to avoid the potential animosity, high legal fees, lost time, and focus that fighting for the name in court would have generated. Today, when people ask us what AACE stands for, we simply respond – "African ACE" - as in the game of cards, where the Ace has the most value!
>
> **Quick question:** If you already have a company name, what does it symbolize, and do you have a registered trademark?

Designing a logo and selecting brand colors: There are many free resources available to support you to create your company's logo and determine what colors to select.

In terms of simple pointers to consider when selecting colors for your company, consider these questions.

- What colors resonate with you?
- What colors resonate with your potential customers?
- What do the colors say about your product or service?
- How differentiated are these colors from the existing products or services in the market ?

Brands in the agriculture, food, and nutrition sectors often use green, brown, and orange for their logos. These colors define the logos for HelloTractor, Garden of Coffee, IITA, African Green Revolution Forum

(AGRF), and Good Nature Agro. We chose the green and orange colors for AACE Foods, primarily because of our mission, rooted in agriculture and local sourcing, and the productivity of the orange fruit on the tree. Canva's color palette provides useful insights to guide your color selection process.[1]

Tech companies such as Skype, IBM, Twitter, and Facebook all have blue logos. Blue reflects trust, serenity, and tranquility. Black is a traditional color of grief and mourning in many countries across the world. As a result, it is hardly ever used for healthcare, baby care, family products, food, or finance.

Developing a tagline: A tagline is simply a description of your value proposition – what benefits you are creating in the marketplace from the customer or client perspective. It is important to note that your tagline can change based on the evolving brand identity and value proposition.

When we first established AACE Foods, our tagline was "Adding sweetness & spice to life." At that time, our core products were jams, chili pepper sauce, spices, and seasonings. Today, the tagline is "Pure Taste," which is linked to our commitment to local sourcing to ensure traceability and our use of natural and healthy ingredients. Other examples of taglines utilized by food and agriculture companies across Africa include the following:

- CowTribe: Healthy Animals, Happy Farmers
- Glow Healthy Smoothies and Snacks: Fresh Raw Material
- Java Foods: Quick, eeZee, Nutritious
- One Acre Fund: Farmers First
- Reelfruit: Delight in Every Bite
- Twiga: Revolutionizing African Retail

Your brand promise: At its core, your brand promise expresses what customers or clients should expect from a product or service in terms of standards and impact. This can be described as a brand's value proposition, which can be used to differentiate your offerings from others. For example, according to Baby Grubz,[2] a Nigerian company,

> All our children's food products address moderate malnutrition and micronutrient deficiencies. All our products are made in a hygienic facility, and we promise to use only the freshest natural ingredients for our baby food. We will entertain no chemicals in our infant's

tender tummies. All children should have access to optimum nutrition regardless of the depth of their parent's pockets.

In defining a company's brand promise, it is essential to clearly outline its values, which encapsulate your brand's personality, purpose, and positioning. Without core values, your brand identity risks blending into the broader ecosystem of companies offering similar goods/services.

In addition, explicitly outline your brand positioning, which directly informs your marketing strategies. To position your brand, you need to be aware of what makes your brand unique and the similarities and differences that your brand shares with others. Once this has been done, brand positioning signals to consumers why they should choose your brand over others and informs how consumers perceive your brand. For example, Egypt's Glow Healthy Smoothies and Snacks' brand messaging calls out why it is different:

> We believe that healthy should taste good and feel good too. A healthy and balanced lifestyle should be affordable and accessible for anyone who desires to embrace it. Our handcrafted juices, smoothies, and acai bowls are natural, clean, and always buzzing with the most energizing and nutrient-rich ingredients.[3]

You may feel that you completely understand your brand and its power; however, it is ultimately, consumers that define this based on their experiences and their understanding of your brand. This concept – brand perception – is instrumental because it is a good measure of how effective brand and marketing strategies are. You must develop a cost-effective and straightforward approach for determining your brand perception annually.

Box 5.2: MANELI PETS' BRAND STRATEGY

An example of a company with a brand that encompasses these core features is Maneli Pets, a pet food company based in South Africa. The Maneli Pets brand has an explicit promise, values, and precise positioning. While perception is hard to measure, founder Nhlanhla Dlamini has been able to track how different markets respond to his branding.

> **Brand promise:** Maneli Pets promises customers excellent quality human-grade pet food, a mix of sustainably sourced wild proteins (e.g., ostrich, crocodile, and venison) and free-range chicken, beef, lamb, and pork. At the core of their promise is sustainably sourced meat processed in world-class facilities.
>
> **Brand values:** Maneli Pets has two core values: sustainability and quality. Every aspect of the company's website and social media speak to these values. Motivated by their values of quality and durability, the brand promises only the highest quality meat is used for their pet food and guarantees that all meat is sustainably sourced. The brand's focus on sustainability shapes the language used throughout the website.
>
> **Brand positioning:** Founder, Nhlanhla Dlamini, validated research that revealed that pet owners are more likely to spend more on pet food than on feeding themselves. This insight into purchasing patterns informed his brand positioning. The value-add of buying from Maneli Pets is that customers know that purchasing this pet food brand does not negatively impact the environment and supports communities in South Africa.
>
> **Brand perception:** Brand perception is hard to measure because it is something experienced by consumers. However, in an interview with Dlamini, he explained that Maneli Pets has to stress different aspects of its brand to appeal to different markets. For example, in South Africa, they stress the high-quality nature of their product because this appeals to South African customers. In Asian markets, the company highlights the exotic quality of its meats because this appeals to Asian consumers.
>
> It is important to note that the Maneli Pets website outlines the brand's promise and values and, in doing this, communicates the Maneli Pets brand to all site visitors. Effective branding needs to be clear on your company's website and social media.

The branding of commodities is difficult but equally essential. Consider WIN Industries, a Senegalese water bottling company that has been able to differentiate itself from other bottled water brands by positioning itself as affordable and responsive to consumers' needs. By introducing a 0.3L bottle, a large segment of the population who purchased water in large sachets switched to bottled water. While the product itself can be considered similar to competitor bottled water brands, WIN Industries has been able to position itself as innovative.

Your marketing strategy

As outlined in Chapter 2, from the onset, you must identify your primary target customer to understand the best way to communicate. Many early-stage entrepreneurs assume that everyone is or could be their target customer. Sadly, this perspective will lead you down a very uncoordinated and often wasteful approach to building your brand. In understanding your target customer, you must identify their values and interests, demographics – age ranges, income levels, where they live, buying habits, preferences, and how much time they spend on various social media platforms.

Customer segmentation is usually based on critical groupings, which may be distinct or overlap.

- **Geographic segmentation:** By country, region, city, or another geographic basis.
- **Demographic segmentation:** Based on identifiable population characteristics, such as age, occupation, marital status, etc.
- **Psychographic segmentation:** Involves an understanding of a consumer's lifestyle, interests, and opinions.
- **Benefit segmentation:** Specific benefits they are seeking from the product, such as convenience, status, value, health, and environmental considerations, etc.
- **Behavioral segmentation:** Based on their relationship with the product or the firm, heavy or light users, brand loyal or brand switchers.

Once you have clearly defined who your customer is, it is critical that you develop precise alignment of the six Ps[4] of your marketing strategy – product, price, packaging, place, promotion, and people – to determine what tactics and channels that you can leverage.

Product: This describes your product or service and what you plan on marketing to customers and consumers, as already discussed in Chapter 2.

Price: In deciding your pricing strategy, while variables like cost are essential, you must consider your target customer, purchasing power, and desire for status. There are three standard categorizations of food products based on pricing: value products, often viewed as the most affordable products in the market; mainstream products that are moderately-priced within their category; and premium products at the higher end of the market's pricing scale. Companies such as Dufil introduced Indomie, an instant noodle product, as a value product, keeping its price point at below $0.20, pegged to the price of one egg, to appeal to the masses of people in Nigeria.

Packaging: Packaging is often a customer's first interaction with your product, and should, therefore, be a physical representation of your brand. As a result, the consumer should instantly recognize new products as belonging to your brand. Indeed, the most successful companies can marry practicality with creativity to use their packaging to attract potential customers.

Choosing the range of sizing options varies by product type and target market. It is also essential to consider how individual packs will fit into bigger containers for ease of transportation, storage, and display on shop shelves. Some products lend themselves to larger packaging sizes, as customers gain more utility from buying the good in bulk. For example, Unga Group Ltd., a Kenyan food processing company specializes in maize flour, primarily used to make ugali, one of the nation's staple dishes. Families make large portions of ugali multiple times a week; therefore, Unga packages its flour in 2kg and 4kg packs.

In contrast, Tropical Heat, another Kenya brand, produces spices, seasonings, salts, and herbs. The average consumer uses a small amount of product per week, which lends itself to 50gram packs. However, Tropical Heat capitalizes on the size of its product mix by offering promotional "mix packs," bundles of 3, 6, and 9 different spices that can be bought together for a discounted unit price. This strategy directly meets consumers' needs by allowing them to select bundles specific to whatever mix they need for their cooking. The company takes this one step further by including one sample recipe on the packaging for each spice bundle that uses the seasonings in the pack and a link to their website for more recipe ideas. This is an effective way to raise product awareness and attract first-time buyers and a fantastic example of using packaging as an advertising medium.

In terms of packaging material, functionality and cost are often the most critical considerations. Walking through your product's user experience is a quick way to assess all possible opportunities for improved functionality and ease of use. Will the material need to be refrigerated or kept in the freezer? Does it need to be microwave-safe? Is it appropriate for long-term use? Is it easy to open? Should it be easily resealable? Or waterproof? Does it change color with long-term sunlight exposure in open-air markets? Considering these questions will help you to select a material that fits all of the customer experience requirements.

In recent years, environmental concerns have led to more and more companies rejecting traditional, bulky packaging in favor of ecologically friendly alternatives. This could include using less packaging or switching to reusable or recyclable packaging material.

Place: This refers to where consumers can access your product or service. You need to ensure that your product and services are accessible physically and virtually to reach your target audience.

Promotion: This simply refers to the strategies you use to build awareness about your product and service, including advertising and public relations strategies.

People: This refers to the individuals representing the business, often centered around the role of the founder and the sales and marketing team members in attracting and retaining customers. It is essential to consider how you want customers to view your employees.

You must ensure alignment between the six Ps of your marketing strategy. This essentially means that there has to be coherence and that each component should support or reinforce the other. For example, you cannot launch a product that appeals to the masses of people but position it as a premium product and put it on the shelves of supermarkets that cater to the upper-middle-class populations.

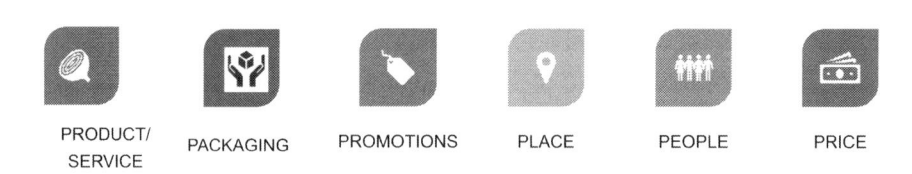

PRODUCT/SERVICE   PACKAGING   PROMOTIONS   PLACE   PEOPLE   PRICE

Figure 5.1 The six Ps.

> **WORKSHEET 5.1: CLARIFYING YOUR BRANDING POSITIONING**
>
> - What is your brand story? Think about the narrative you want to portray to consumers, clients, suppliers, employees, investors, and other stakeholders.
> - How does your story compare with others? How differentiated is your story?
> - Who is your target audience? You need to think about who you are telling your story to, as this will inform your language and your marketing strategy.
> - What makes your brand unique? Think about why a consumer or client should choose your product or service over others.
> - What are your brand values?
> - How do your brand's values inform your messaging?
> - How do you want your customers to feel about your product or service?
> - Is there alignment between the six Ps of your marketing strategy? If not, what needs to change?

Leveraging more traditional marketing approaches

As an entrepreneur, determining which strategies to leverage will vary based on the product or service that you offer, your customer segment, and your marketing/branding budget. There are a range of options to consider.

Finding a credible spokesperson: Most of the founders interviewed as part of the research for this book served as brand ambassadors for their companies. They were able and willing to tell their brand story and to make themselves public figures. These founders are either active on social media or can attract local or international press coverage. For example, Jehiel Oliver of Hello Tractor specifically enrolled in a communications course and engaged a speech specialist to build his skills and capabilities. Early on in his company's journey, he had the unique opportunity to speak at the Global Entrepreneurship Summit in Kenya with U.S. President Barack Obama. He leveraged this and other forums to promote Hello Tractor. Jehiel spends a significant amount of his time listening to podcasts by CEOs of multinational companies and reviewing earnings reports to enable him to engage with partner companies effectively and build relationships.

If you would prefer to be in the background, it is crucial that you identify another team member, customer, partner, or even engage a credible external voice that can serve as your spokesperson.

Champions: It is useful to have champions for your brand to bridge the credibility and trust gap for investors and partner organizations. Consider the experience of Hadija Jabiry, the founder of EatFresh, a Tanzanian food company that specializes in growing, processing, packaging, and selling high-value fruits and vegetables locally and internationally. When Jabiry started the company, she had already failed at previous enterprises and did not have any funding to build her brand locally. She reached out to a few television stations to invite them to visit her farm, but they were unresponsive or too expensive. She eventually persuaded senior officials from the Ministry of Agriculture to visit her operations. Being top state officials, they came along with a press crew, and the visit was aired on major national TV stations and in the print media at no cost to her company. This breakthrough attracted significant interest from funders, customers, and potential partners from which the company continues to benefit.

Networks: Building strong national and international networks is essential. A clear example of the power of strong networks was reinforced by Dlamini of Maneli Pets, who has leveraged the networks that he built during his educational and work experiences to promote his business. He has benefited from referrals to global customers and investors. The initial funding for the establishment of his pet food factory resulted from a business school network.

Radio: Leveraging radio is often considered the most effective marketing tool when there is a need to reach large audiences over a wide coverage area to create top of mind awareness about a brand or product. It is estimated that at least 75% of households in developing countries having access to a radio. As an entrepreneur, you must think creatively about developing a crisp and clear message in the relevant languages and selecting the most appropriate radio station and a time slot that fits your client or customer's listening habits. This is especially important when targeting farmers. As Sahel Consulting's Associate Partner Temi Adegoroye underscores, farm radio stations are ideal for marketing agricultural related products such as fertilizer, seeds, equipment, and logistics or mechanization services. However, it is imperative to understand the listening behavior of farmers and align the deployment of campaigns and adverts accordingly. According to Adegoroye, "Most farmers listen to the radio when they are not on the farms – early in the morning or late in the evening. Any adverts deployed and programs aired outside of these periods may not reach the farmers and may be ineffective."

If you have a limited marketing budget, actively seek out opportunities to serve as a guest for a popular interactive radio show. You will have a chance to share insights on your company, products, or services. Over 5 years, I anchored a popular radio show called "Fidelity SME Forum," which aired from 6.30 a.m. to 7 a.m. every Tuesday. This time slot was perfect for commuters en route to work in Nigeria's major cities. I featured numerous entrepreneurs in the food and agriculture landscape who were the bank customers. From anchoring this show, I learned the power of free advertising, while also selecting the most appropriate station that appeals to your target audience and the time slot that would enable maximum reach.

Print media: This includes articles, features, and advertisements in newspapers, magazines, brochures, flyers, and even books. Similar to the radio, the effectiveness and suitability of this type of media is hinged on your target audience and how this type of approach appeals to them. Newspaper coverage can often be obtained at minimal or no cost and often appeals to corporate partners and international development organizations. Still, this approach would typically not reach farmers or grassroots stakeholders. Printed flyers with the right aesthetic qualities of colors, images, and emotional content may be more appealing to farmers, especially if the messages are simple and straightforward.

Branded gifts: Branded gifts are giveaways to reinforce a name, logo, and a message and ensure a longer-term connection with the customer or influencer. These tools are often used in fieldwork to refresh farmers' memory of the uniqueness of a product's brand and critical attributes. Branded gifts also enhance the penetration of marketing messages and brand expansion because gift holders serve as active influencers and promoters. Via the Yam Improvement for Income and Food Security in West Africa II project, which focused on promoting yam seed systems in Ghana and Nigeria, Sahel Consulting designed branded gift items. These T-shirts and caps were distributed to farmers during field day events to promote the activities of the project, private seed company partners, and create awareness on improved yam varieties.

Farmer field days: Farmer field days are the most effective method of introducing new agricultural products, innovations, and practices to smallholder farmers. On a field day, farmers in communities in close proximity are gathered at a demonstration site where training and discussions are held to promote new agricultural practices and improved inputs such as seeds, fertilizers, and crop protection products. A field day is successful when farmers leave with knowledge and insights about what is possible after experiencing what other farmers have accomplished using improved practices. Under the Building an Economically Sustainable Integrated Seed System for Cassava (BASICS), Sahel Consulting, working through Context Development Network, in collaboration with the International Institute of Tropical Agriculture (IITA) and other stakeholders, utilized farmer field days to promote the adoption of improved cassava varieties among farmers.

In BASICS, Demand Creation Trials (DCTs) were established on the fields of integrated cassava processors in Nigeria to demonstrate the performance of more than ten new and improved cassava varieties to farmers. During the field days, each improved cassava variety was compared with the farmers' local best type. Also, farmers were able to select the best performing of the new varieties on the demonstration plot. In some cases, farmers were given cassava stems of the improved varieties to try on their farms.

Partnerships/Bundling: Many new entrants have benefited from partnering with established brands to bundle their product offerings and expand their reach. For example, Nutzy Peanut Butter, when trying to enter the Nigerian market, bundled its products with established bread brands, to induce trial. Some wine and cheese companies have partnered to host joint promotional events.

Partnerships are also an excellent way to enter a new market, leveraging the manufacturing capacity, brand awareness, and distribution networks of an entrenched player. For example, Rwanda-based Africa Improved Foods, partnered with Tropikal Brands in 2019 to introduce its Nutripro[5], a nutritious porridge flour into Kenya. Tropikal Brands also has similar manufacturing and distribution partnerships with Danone and Associated British Foods (ABF) focused on the Kenyan Market.

Roadshows/Market storms: Given that open-air markets still dominate many African countries, market storms are actively utilized by food processors and fast-moving consumer goods (FMCGs) to generate interest and excitement about the products. This describes a coordinated set of activities when five to ten sales team members and paid promoters enter a market for a day or series of days, depending on the size of the market, with the same branding materials, leveraging T-shirts and hats. They overwhelm the space with their visible presence and engage in games with attractive product prizes and sampling. Using music and dancers in a moving bus or truck, they drive multiple and intense engagements to attract potential customers and generate interest in the product.

Companies that target children and youth often engage in school visits, sponsor interhouse sports, or partner with schools to host tailored events.

In-store tasting, promotions, and displays: People typically enjoy winning something at no cost to them. Some food processors use spinning wheels or peel-off stickers to determine who wins what, generally linked to luck. Contests offer an attractive marketing vehicle for small businesses to acquire new clients and create awareness. Other companies offer free samples or tasting, setting up stands at a cost within or right outside retail outlets. Providing a small taste usually induces a large purchase. Others offer free recharge cards or airtime, linked with use.

In addition, consider opening small sampling stalls in markets and outlets around your city. This is a low-cost mechanism for potential customers to taste your product, ask questions, and even get to know your staff personally, all of which provides incentives for them to make purchases on their next shopping trip.

While still uncommon in many African markets, coupons provide an affordable marketing strategy for small businesses, drawing first-time customers with discounts linked to purchase amounts or a flat percentage offered on a sale.

Route to market (RTM): This strategy is a step-by-step plan that details how firms get their products from manufacturing to consumers. Depending on the product and market you are working with, your RTM strategy can vary, so it is essential to consider mapping out the process.

Central to any RTM strategy is your target market. Understanding your average consumer can give insights into which sales channels will be most impactful, allowing you to settle on a plan that matches your budget and goals. Examples of sales channels include personal selling, e-commerce, retail, wholesale and direct marketing. Traditional FMCG companies typically use the wholesale channel to get their products to market. For example, Brookside Ltd. partners with more than five of Kenya's largest retailers to deliver their product to consumers.

On the contrary, young enterprises sometimes engage in direct and relationship marketing. Sometimes known as BTL (below the line) marketing or experiential marketing, this channel relies on direct communication with potential consumers, using initial customer satisfaction to leverage sales. This can include samplings, events, market storms, visibility advertising, point of sales promotion, direct marketing, and telemarketing.

The case of Java Food's eeZee Noodles, Zambia's first indigenous noodle brand, highlights the importance of marketing. Monica Musonda, the founder of Java Foods Limited, was inspired by the popularity of noodles in Nigeria. She wanted to replicate this in Zambia. Musonda faced a significant problem: a large population of Zambian consumers were not familiar with the product. Existing noodle brands were marketed to expatriates and foreigners. Musonda found that the answer to this problem was effective marketing. As the price of bread soared, noodles became an attractive staple. Circumstance, coupled with digital marketing through social media, the use of local languages, and brand activation, propelled eeZee Noodles to become Zambia's most popular instant noodle brand. Marketing spread awareness of eeZee

Noodles and created demand where there was none. EeZee Noodles' effective marketing created a new staple for Zambian consumers and promoted the eeZee Noodle brand.

> **Box 5.3: CASE STUDY – BROOKSIDE DAIRY LTD, KENYA[6]**
>
> Brookside Dairy Ltd. is one of Kenya's most recognizable indigenous food brands. Founded in 1993, the dairy processing company has sustained a market share of 45% since 2016, thanks to its diversified distribution strategy and strong brand identity.
>
> With a wide product mix of milk, cream, butter, and yogurt, it was essential that Brookside unified their commodities under a distinctive, targeted, and consistent brand image. The company's target market consists mainly of heads of households (generally mothers shopping for their families). Brookside strove to create a family-friendly image that could appeal to consumers of all ages. The company utilizes a simple logo of green pastures against a blue sky and the slogan "Goodness for All" to reinforce a message that their products are natural/organic, without tying their brand image to a particular demographic. The simplicity of their imagery and slogan make their brand easily recognizable on supermarket shelves.
>
> Brookside has utilized digital media to further its brand's strength and reach a variety of consumer demographics. Its website is regularly updated to position the brand as the "healthiest option" by detailing recent investments in product quality. As a Kenyan brand with ties to the nation's founders, Brookside emphasizes its identity as a brand created for Kenyans by Kenyans by highlighting a commitment to social impact on the website's "impact page."
>
> To consolidate its market share, Brookside uses different distribution platforms to boost sales of its products. It currently sells to outlets in countries around East Africa through agents, subagents, and distribution depots. To acquire these distribution channels, Brookside prioritized creating a sharp brand image across the region to inspire confidence in potential distribution partners. The team did this through investing in radio and television advertising in neighboring countries, direct outreach to distributors, and high-profile collaborations to raise brand awareness, such as Walt Disney's licensing of iconic cartoon imagery to the company in 2017.

Your online presence

Every company needs an online presence, but the depth and breadth depends on the type of product, service, and client/customer engagement required. At the very least, a company should have a landing page that briefly outlines its mission, vision, and value proposition. However, most companies should aim to have a robust and dynamic website and an active presence on Twitter, Instagram, Facebook, LinkedIn, and YouTube. These are affordable and accessible media channels for attracting, retaining, and engaging customers. Leveraging social media and recruiting influencers and bloggers to promote your product on their platforms is often an effective method.

Consider the example of Glow Healthy Smoothies and Snacks based in Cairo, Egypt. Given the global move toward healthier lifestyles and the rise of alternative diets, the brand aims to be relevant and contemporary and appeal to different demographics. Glow's presence is mostly felt online through its website, Facebook, Twitter, and Instagram pages. Amina Rashad, the company's founder, explains that because Glow does not have a physical location, maximizing its presence online has been critical to its success. Glow uses content marketing for advertising its products and informing customers on where to find them. The company offers products via pop-ups around Cairo and delivery services. The content on its online platforms is updated daily and highlights a vibrant, healthy, and mobile customer. The company also interacts with its customers via an e-newsletter and a weight-loss program. Glow can catch the attention of consumers by appealing to their need for healthy options on-the-go. The company cultivates interest by highlighting the high-quality ingredients used as well as the preparation process. Desire is created through their strategic use of pop-up locations where they increase their physical presence. The company also creates demand by heavily using language associated with health, weight-loss and wellness. Lastly, they inspire their customers to act by ensuring accessibility.

When using online tools, consider the AIDAR model[7], which outlines five critical components of your messaging. First, your ability to get the attention of potential customers by standing out from the noise in the social media channels. Second is to engage them and build their interest by showing your potential customers your value and how you are different from your competitors. Third is generate enough interest that they actually demand your product/service. Fourth is that they take action and make a purchase. Fifth is that you retain them by providing opportunities for feedback and follow-up to build loyalty and a repeat purchase or a referral.

Digital marketing: An excellent approach to help you find the best digital marketing channels for your business is to think of your marketing channels in three buckets: owned, earned, and paid.

Owned media channels: These are channels you have full control over, such as your website, social media accounts, and mailing list. From a small- or medium-size enterprise's (SME's) perspective, a website is a beneficial resource. It gives you a medium through which to share your brand and story and can also shape perceptions of your brand and raise awareness. For example, Twiga's website clearly outlines its mission, vision, and values and highlights its successes. This shapes the way consumers and funders perceive the brand.

Paid media channels: These are any form of advertising and marketing for which you pay. This includes adverts on social media sites, paid features, and Google AdWords. While paid media is essential, you have to be cautious not to invest in marketing that does not translate into increased awareness and revenue. This is where constant feedback, analytics, and evaluations are useful, which allows you to understand what works and what does not.

Earned media channels: Any external means through which your company gains exposure. This allows you to raise awareness and promote your business through outside channels and will enable you to leverage the reach of these external platforms. An excellent example of a company that does this brilliantly is Hello Tractor, which has built strong international recognition by being featured in *Forbes, Fast Company, Washington Post*, and other leading international online and print media. By building strategic relationships with key multinationals engaged in tractor production, the company has benefited from their placements, reach, and influence.

According to the digital media trifecta model[8], the best combination follows a 2:1:1 ratio – two owned media channels, one paid media channel, and one earned media channel.

Table 5.1 Examples of African Agribusinesses and Their Social Media Footprint

Company Name	Tagline	Website	Twitter	Instagram	LinkedIn	Facebook	YouTube
AACE Foods	Pure Taste	https://www.aacefoods.com	@aacefoods	Aacefoods_ng	aace-food-processing-and-distribution-limited/	aacefoods.com	
Babban Gona	Better your life	https://babbangona.com/	babbangona	babban.gona/	babban-gona/	OfficialBabbanGona	
blueMoon	"Once in a blue moon, exceptional ideas turn into great companies that change the world"	www.bluemoonethiopia.com	bluemooneth			blueMoonEthiopia	
Farmcrowdy	Empowering Farmers. Together	https://www.farmcrowdy.com/	farmcrowdy	farmcrowdyng/	company/farmcrowdy/	farmcrowdy	YouTube: https://www.youtube.com/channel/UC6pFNkiAvTIUbRG8v-Js8gQ
Garden of Coffee	A coffee brand millennia in the making	https://gardenofcoffee.com/	gardenofcoffee	gardenofcoffee		gardenofcoffee	
GBRI/EatFresh	Enjoy a Healthy Living	https://eatfresh.co.tz/about-us/		eatfreshtanzania		EatFresh-Tanzania-971806219658714	

BUILDING YOUR BRAND 103

Table 5.1 Continued

Company Name	Tagline	Website	Twitter	Instagram	LinkedIn	Facebook	YouTube
HelloTractor	Connecting you to your tractor and your tractor to the world	https://www.hellotractor.com/home	HelloTractor	hellotractor	https://www.google.com/url?q=https%3A%2F%2Fwww.linkedin.com%2Fcompany%2Fhello-tractor%3Ftrk%3Dpro_cprof&sa=D&sntz=1&usg=AFQjCNG9SMSkIvIdeTPUXuHeqauMgAK1Lg	hellotractor	
Tula	Market Access for Smallholder Farmers in Rural Africa	https://www.tulaa.io/	https://twitter.com/tulaanews		https://www.linkedin.com/company-beta/11164051	https://web.facebook.com/tulaanews?_rdc=1&_rdr	YouTube: https://www.youtube.com/user/esokonetworks

Social media strategy: Developing your social media strategy requires four key steps. First, you will have to define your goals clearly. In the case of startups and niche products, your social media marketing strategy may begin with the need to test ideas, create awareness, and build anticipation for new products and services. In other cases, goals may be more specific – boosting sales, geographic expansion, increasing real-time brand engagement, or generating quality sales leads. Second, you have to determine your target platforms based on your understanding of your target customer's identity and preferences. Different social media platforms appeal to different demographics, and it is crucial to find out where your audience spends time online. For example, younger audiences may be more effectively reached on newer platforms, such as TikTok. Brands with an emphasis on aesthetics may want to develop a more visual strategy using Instagram. Geography is also a key factor; WhatsApp and Twitter are popular in some countries, while Facebook is more prevalent in others. Third, it is vital to develop a clear content strategy that outlines the types of content you will share, ensuring that it was unique, useful, and shareable. For example, articles and blogs may be suitable for one audience, while videos and pictures will drive engagement with another customer group. In addition, some content is more suitable for one channel than others. Fourth, determine how to ensure a dynamic and vibrant approach that allows for active engagement. This will ensure that your strategy is actively listening and engaging with questions, concerns, and suggestions and receiving and acting on feedback.

Attention	Interest	Desire	Action	Retention
You need to stand out to attract potential consumers	Show customers your value-add and how you are different from your competitors	Create demand for your product/ service.	Inspire consumers to take a specific action and make a purchase	Provide opportunities for feedback, follow up and support to build loyalty

Figure 5.2 AIDAR model. Source: Smart Insights. The AIDA Model. Accessed: https://www.smartinsights.com/traffic-building-strategy/offer-and-message-development/aida-model/ (1 August 2020).

> **Box 5.4: TWO DISTINCT APPROACHES TO BRAND BUILDING**
>
> **The case for physical marketing**
>
> Digital marketing is vital; however, for a few entrepreneurs, digital marketing has proven ineffective. For Maneli Pets and Esoko, introducing customers to their products physically has been the winning strategy. Maneli Pets found that digital marketing did not bridge the trust gap between its products and its customers. Going to trade fairs and exhibitions, developing physical stores, and offering testers proved to be a far more effective way of marketing its products than the expensive digital marketing strategy in which the team had previously invested. Similarly, Esoko found that organizing open houses in different towns created awareness and translated to an increased customer base.
>
> **The case for digital marketing**
>
> Digital marketing is relatively cheap, measurable, and can be tailored to reach particular demographics. For both startups and larger companies, digital marketing has been incredibly successful. For Local Village Foods, a South Africa food production startup, content-based marketing on Instagram has helped the company build a brand and a community. Similarly, Olam, a much larger company, has found digital marketing extremely useful. While the company does not sell directly to consumers, it uses social media to raise awareness about its initiatives and establish itself as leader in the ecosystem.

Introducing unusual products and dealing with crises

Reputation management, often referred to as public relations (PR), is critical in the life cycle of a brand, and every entrepreneur must actively develop strategies to protect it.

One company that has been able to build a formidable brand in record time is the Tolaram Group, with its product, Indomie. Before the company entered Nigeria, the average consumer resisted any noodle products, citing a strong resemblance to worms. Noodles were alien to the Nigerian diet and were viewed as unappetizing and unappealing.

In 1988, drawn to the large and growing population in Nigeria and rapid urbanization, the Aswani brothers,[9] operating out of Indonesia, decided to introduce their noodle brand into the country. The first few years proved extremely difficult, as they embarked on widespread customer engagement and education. They also lobbied nutrition organizations to endorse their product. While trust was being built among consumers, especially children, they had a significant setback. In 2004, the National Agency for Food and Drug Administration and Control (NAFDAC)[10], confirmed that three batches of Indomie Noodles were contaminated with carbofuran, a carbamate pesticide used in agriculture.

There were rumors that children had died from consuming these contaminated batches. In a swift response, Tolaram launched a massive campaign to rebuild trust. They introduced a catchy song on all radio stations, focused on families, with a brand promise on product quality and integrity. Children who could call in and sing the entire song during the live radio show would win boxes of products and a visit to their factory to verify the brand promise. They also launched school remodeling programs and an annual Independence Day Awards for Heroes of Nigeria to celebrate exemplary Nigerian children who had overcome severe obstacles or engaged in heroic acts.

In 2020,[11] "the brand enjoys near-universal name recognition in the country, maintains a 150,000-member fan club with branches in more than 3,000 primary schools."

Financing your branding and marketing efforts

For SMEs, finding the funds for their branding and marketing strategies is often a challenge.

Where should the money come from? For established firms, the marketing budget comes from their revenue. However, newer businesses may need to spend more on marketing. In this case, the marketing budget should be based on the firm's projected revenue and often ranges from 8% to 20% of its budget.

To justify any allocation to a branding or marketing initiative, the team must submit a clear business case, which addresses a range of critical questions including:
- What is the opportunity?
- Who is the target market?
- What is the strategic value of the opportunity for the company?
- What is the projected revenue and profit potential?

- What is the time to generate the revenue?
- What is the impact on revenue and sales capacity?
- What other opportunities will this impact?
- What are the risks?
- What are the implications of passing on the opportunity?

How should it be allocated? Treat marketing like a small business unit. Budgeting your marketing spending as an ongoing monthly cost helps ensure that your efforts remain both manageable and consistent.

For each marketing/distribution tactic utilized, how do you measure impact? There are analytical tools that allow for relatively easy, yet effective tracking. It is essential to know what works and what does not and whether or not you are communicating the essence of your brand.

Table 5.2 Measuring the Effectiveness of Your Marketing Strategy

Tactic	Potential Impact Measurement Strategy
• Buy-one, get one free. • Bundled, get a free milk sachet, when you purchase a tin of milk. • Loyalty rewards scheme.	Sales volume; consumer satisfaction.
• Win a prize when you purchase a product valued at a certain amount. • Entered into a drawing when you purchase a product/basket of goods valued at a certain amount.	Price of average customer's basket of goods; sales volume.
• Partner with companies selling products for which demand is complimentary with your goods – buying goods from your company gives consumer free or discounted goods from the other.	Sales volume across both companies; product awareness.
• Temporary price reduction.	Sales volume; market share; product awareness.
• Price-packs ("X% extra free"). • Competitor price matching offer.	Sales volume; market share.
• "Last chance to buy" offers.	Sales volume; percentage of first-time buyers.
• Free recipes included in purchase.	Customer satisfaction; product awareness.
• Free sampling.	Sales volume; product awareness.
• Trial period and money-back guarantee.	Product awareness; initial sales volume (volume of product sold); sustained sales volume (amount of product sold and not returned); percentage of first-time buyers.

What software is available? How can entrepreneurs leverage it?
- Most social media sites track their data and offer their analytics services for free. Instagram has some third-party apps and agencies can often give more profound insights than are available on the social media site
- Digital branding allows entrepreneurs to receive instant/real-time feedback on how successful their campaign is. There are no clear-cut approaches for measuring impact; however, consider experimenting with a few of the following strategies:
    - Clickthrough rates
    - Conversion rates
    - Reach
    - Impressions
    - Profile visits
    - Audience location and demographics

Summary

Building a brand requires a concerted effort – first to define the image you want to create in the public domain, then to identify and utilize appropriate tools and tactics to convey a consistent message, deliver impact to authenticate this image and leverage strategic approaches and partnerships to build and sustain trust and credibility.

Beyond the 6 Ps (product, price, packaging, promotions, place, and people), entrepreneurs have to embrace traditional and social media, investing in a toolbox of tactics and strategies that effectively reach their target customers while actively monitoring impact and responding appropriately.

Ultimately, your ability to build a strong brand requires courage, consistency, coherence, and patience to set your company apart and amplify its voice, products, and services first locally and, ultimately, for a global audience.

Notes

1. Canva. (n.d.) *How to choose the right colors for your brand* [Online]. Available at: https://www.canva.com/learn/choose-right-colors-brand/ (Accessed: 1 August 2020).
2. Baby Grubz. (n.d.) *Our promise* [Online]. Available at: https://babygrubz.com/about-us/our-promise/ (Accessed: 30 August 2020)

3 Glow. (n.d.) *Home page* [Online]. Available at: https://www.getyourglow.me/en_US (Accessed: 1 August 2020).
4 Rakhareve, K. (2016) "Why you need the 5 p's of marketing," *Dragon Fly Marketing*, 2 November [Online]. Available at: https://dragonflymarketing.co.za/blog/5-ps-marketing/ (Accessed: 11 August 2020).
5 Tropikal Brands. (n.d.) *Nutripro family porridge flour* [Online]. Available at: http://www.tropikal.co.ke/project-details/nutripro-family-porridge-flour/ (Accessed: 1 August 2020).
6 Brookside Dairy Limited. (n.d.) *Home page* [Online]. Available at: https://www.brookside.co.ke/ (Accessed: 1 August 2020).
7 Hanlon, A. (2020). "The AIDA model," *Smart Insights*, 10 September [Online]. Available at: https://www.smartinsights.com/traffic-building-strategy/offer-and-message-development/aida-model/ (Accessed: 1 August 2020).
8 Springer, J. (2015) "The digital marketing trifecta," *JRS Strategies Blog*, 30 May [Online]. Available at: https://www.jrsstrategies.com/the_digital_marketing_trifecta (Accessed: 1 August 2020).
9 Buchanan, L. (2019) "How 1 company--and its insanely popular and cheap noodles--transformed Nigeria: Disruptive innovation guru Clayton Christensen explains why entrepreneurs are our best hope to eradicate poverty," *Inc.*, 11 January [Online]. Available at: https://www.inc.com/leigh-buchanan/clayton-christensen-prosperity-paradox.html (Accessed: 1 August 2020).
10 Ahiante, A., Osagie, C., Ibaim, A., and Momodu, S. (2004). "Nigeria: NAFDAC recalls Indomie for destruction," All Africa, 20 May [Online]. Available at: https://allafrica.com/stories/200405200336.html (Accessed: 1 August 2020).
11 Buchanan, L. (2019) "How 1 company--and its insanely popular and cheap noodles--transformed Nigeria: Disruptive innovation guru Clayton Christensen explains why entrepreneurs are our best hope to eradicate poverty," *Inc.*, 11 January [Online]. Available at: https://www.inc.com/leigh-buchanan/clayton-christensen-prosperity-paradox.html (Accessed: 1 August 2020).

6

FINANCING YOUR GROWTH

Introduction

More than 80% of the entrepreneurs surveyed for this book recognize financing in the food and agriculture landscape as the most significant barrier to scaling on the African continent. These entrepreneurs often have to contend with a few additional hurdles, including establishing linkages and building trust with local and international investors who do not understand the sector and consider it to be highly risky. Financing barriers are more pronounced for women and young people, who often face implicit biases and do not have access to assets that can serve as collateral.

This chapter will provide insights from the experiences of agriculture and food entrepreneurs who have been able to overcome these hurdles and raise significant funds for scaling their companies, leveraging innovative approaches to generating funding internally and externally. It will also provide insights for determining the most appropriate financial instruments and tools for your business, hinged on your business model and life cycle. Finally, it will offer practical steps to ensure investment readiness and financial exits.

Financing options

Your decision process to raise capital should be hinged on your mission, vision, business model, the needs of the business, the cost of the financing, future growth opportunities, and the returns that can be generated from the use of the funds. And, if the offer is for a part of your business (not the entirety), it also depends on who the investor is – their reputation, integrity, knowledge, and the value they can add as a joint owner of the business. There are a range of options available to you, as outlined in Table 6.1, linked to the specific phase of your growth and development.

Startup and early phase

Typically, at this stage, your business is still in its first few months of operations. Your business model is unproven, and sales are unpredictable; they are perceived as risky investments, especially for large ticket funding requirements. During this phase, your company has limited access to collateral, no track record for extensive due diligence, which raises uncertainties about its ability to repay debts and generate returns for investors. As a result, most entrepreneurs will have to rely on their family, friends, angel investors, business plan competitions, challenge funds, crowdfunding, prizes, and fellowships to raise financing.

Table 6.1 Financing Options by the Stage of Your Business Life Cycle

	Startup	Early Stage	Growth	Maturity	Decline or Reinvention
Family/Friends/ Angel Investors	Yes	Yes	Yes		
Business Plan Competitions	Yes	Yes			
Accelerators/Incubators	Yes				
Grants	Yes	Yes	Yes	Yes	Yes
Crowd Funding	Yes	Yes	Yes		
Venture Capital	Yes	Yes			
Challenge Funds	Yes	Yes	Yes	Yes	
Fellowships	Yes	Yes	Yes	Yes	
Impact Investments		Yes	Yes	Yes	
Private Equity Investments			Yes	Yes	Yes
Loans		Yes	Yes	Yes	Yes
Trade Credit		Yes	Yes	Yes	Yes
Payment Advance		Yes	Yes	Yes	Yes
Initial Public Offer			Yes	Yes	

Angels: While friends and family can often act as angels, providing startup funding via loans or equity, there is an emerging formal group of early-stage investors on the African continent who operate via "angel networks." For example, the Lagos Angel Network (LAN)[1] serves as a nonprofit organization that pools seed funders that provide early-stage seed funding and mentoring for startup entrepreneurs. Entrepreneurs are required to provide vital business model information and a high-level pitch via the organization's website before they are invited for a formal session with the angels. The Business Angels select a small group of entrepreneurs to invest in and mentor, while also expecting to earn a healthy return on their investments. The African Angel Business Network (ABAN) is supporting the development of early-stage investor networks. Currently, it includes the Cameroon Angel Network (CAN), Ghana Angel Network (GAIN), and Venture Capital for Africa (VC4Africa), as part of its membership.

It is imperative that you diligently manage your engagement with angels, ensuring adequate documentation, including nondisclosure agreements, clear and decisive contracts, and shareholder documents. In addition, engage in active negotiations to ensure that you obtain the best terms for the investment, and manage expectations to ensure cordial relationships with the investors during and beyond the investment period.

Grants: These primarily refer to funds provided to an entity at no cost to the company and with no expectation that the organization would have to return the funds, unless they are no longer needed. A grant is linked to a direct program or project, and the beneficiary is expected to report on the impact achieved. There are typically two forms of grants: restricted and unrestricted. Restricted grants refer to funds earmarked explicitly for critical activities, which must be accounted for and reported upon. If the funds are not needed for what they were initially requested, they would either be returned to the funder or reallocated to another activity, following a formal request from the grantee. Unrestricted funds are donations that are available for the organization to use for any purpose.

There are a growing number of foundations and initiatives that provide grants to agriculture and food entrepreneurs, especially if you can show a link to social change through interventions, such as innovative farmer engagement or last-mile distribution. Babban Gona has benefited from grants from the Bill & Melinda Gates Foundation, the Mastercard

Foundation,[2] and other funds, given its engagement with farmers and youth in Kaduna State in Nigeria. For example, in 2015,[3] the Gates Foundation gave Babban Gona $4 million to demonstrate the sustainability of an innovative social enterprise to enable smallholder farmers to increase their incomes, improve their livelihoods, and publicly disseminate the lessons learned from such a project.

Investors operating on the African continent include private and corporate foundations, operating and grant-making foundations, and diplomatic offices and development agencies. There are significant variations in the regional focus of many of these funders, with the majority concentrating on specific countries and focusing on select value chains. In addition to the large global foundations, there are a growing number of indigenous personal and corporate foundations established by Africans to support entrepreneurs on the continent.

It is essential to recognize that while the availability of grants is vital for entrepreneurs looking to scale, there are a few critical drawbacks. First, grants can often sidetrack you from your key focus and business model. Donors often provide sizable amounts of funds for predefined programs and projects that may not directly match your company's mission and vision. To avoid this temptation, you must be willing and able to decline funds that do not directly fit your needs. Second, foundations are often reluctant to provide core organizational funding, which covers overhead or administrative expenses. In rare cases, in which they agree to cover these expenses, they are typically unwilling to contribute more than 20% of overall project expenses toward overhead or administrative costs.

Fellowships: These have supported many agriculture and food entrepreneurs on the continent through multisector initiatives that provide startup funding, customized support services, executive education, and advisory services. They include fellowships offered by the Acumen Fund, Ashoka,[9] Echoing Green, Schwab Foundation for Social Entrepreneurship, Skoll, and Draper Richards Kaplan Foundation. Some of these fellowships specifically provide financial support for the entrepreneur for a defined period, covering a portion of their salary to enable the entrepreneur to focus exclusively on implementing their vision and scaling the business.

Accelerators and incubators: A growing number of local and international incubators and accelerators target African entrepreneurs in the food and agriculture landscape, with a preference for companies that leverage innovation and technology. For example, blueMoon[4] is Ethiopia's first youth agribusiness/agritech incubator and seed investment platform. According to its founder and Chief Happiness Officer, Eleni Gabre-Madhin, "Once in a blue moon, exceptional ideas turn into great companies that change the world. We seek to discover, nurture and invest in innovative, scalable, and potentially transformational ideas." Twice a year, blueMoon selects ten startup teams for its 4-month incubator program in Addis Ababa. These teams benefit from skills training, mentoring, a beautifully designed physical workspace with high-speed internet access, and a range of support services, coaching, connections, and seed funding. blueMoon expects the startup entrepreneur to invest personal funds into the business and meet its performance milestones. On its part, it invests $10,000[5] in each startup in exchange for 10% equity. At the end of the incubator program, all startups can access additional funding by participating in a pitch event to the blueMoon Lions, and also benefit from the blueMoon Crowd.

CCHUB[6] in Nigeria has a range of services for startups in diverse sectors, including a 6-month preincubation program to help aspiring entrepreneurs build and launch initial prototypes of their tech-enabled solutions and to validate the ideas and product/market fit. Entrepreneurs also benefit from a $5,000 investment. In addition, CCHUB offers a 12-month incubation program through which startups can access up to $25,000 and have direct access to follow-on funding of up to $250,000 from Growth Capital,[7] free office space for the team, business support, and access to a network of partners. The Hub also offers accelerators that are 12-week programs designed to help startups refine their products, business model, and processes.

In 2019, Village Capital partnered with the Small Foundation and Ceniarth to organize the Agriculture Africa Accelerator.[8] The selected startups included Nigeria's Reelfruit, which makes dried fruit snacks; Uganda's Akellobanker, a startup that allows farmers to purchase inputs on credit; Agro Supply, a mobile layaway service; Kenya's Capture, a workflow streamlining solution; Raino Tech4Impact, a cold chain solutions provider; Senegal's Aywajieune, an online platform for sea products; Ghana's Complete Farmer, an agri-investment solution; Zambia's eMsika, an e-commerce platform; and Rwanda's Volta Irrigation, an irrigation services firm. Two of the companies – Reelfruit and Complete Farmer – eventually received $50,000[9] for their companies.

Visit www.nourishingafrica.com for a list of accelerators and incubators that support agriculture and food entrepreneurs on the African continent.

Crowdfunding: This practice of funding a project or venture by raising financial contributions from many people, typically via the internet, is gaining popularity on the African continent. There are a few types of models for crowdfunding.

- **Reward-based:** The crowd gives money to an individual, project, or business in exchange for a nonfinancial reward. The rewards are generally either shirts or stickers or an early version of a product (mostly, a presale via crowdfunding).
- **Lending-based:** The crowd lends money to an individual or business with expectations of getting the principal back with interest. This is the model utilized by Farmcrowdy in Nigeria.
- **Equity-based:** The crowd invests in a business with hopes of sharing its success as it grows.

Consider the case of Tomato Jos;[10] a startup focused on tomato farming and processing in Northern Nigeria, which raised $50,000 in 2014 from Kickstarter via a compelling video on the high postharvest losses associated with tomatoes and the country's dependency on imported paste. The founders used the donation-based model, offering contributors different gifts, including tomato paste and spices. However, the campaign's exposure also enabled the founders, Mira Mehta and Shane Kiernan, to attract additional funding. As of 2019, Tomato Jos had completed two fundraising rounds and successfully attracted local and international funders.

Competitions, prizes, and awards: Several funding organizations, foundations, and development partners have introduced competitions such as Agribiz4Africa and SheLeadsAfrica to encourage youth entrepreneurs and promote social innovation. In addition, awards such as the World Food Prize Foundation – Borlaug Field Award, the Zayed Sustainability Prize, Anzisha Prize, Africa Food Prize, LEAP Social Innovators Programme & Awards are available for entrepreneurs with funds provided to the winners. While some of these prizes are global, they have started making a concerted effort to incorporate African entrepreneurs with a few explicitly targeting breakthroughs in agriculture and food.

Table 6.2 Examples of Agribusinesses that have Benefited from Competitions, Fellowships, Prizes and Awards[a]

Agribusiness	Awards Won	Description of Award
Abdou Maman, founder and CEO, Tele-irrigation (Niger): Tele-irrigation is a technology process that allows farmers to manage their farms remotely and their irrigation systems, using their mobile phone and solar panels.	Morocco: King Hassan II Great World Water Prize 2015 ($100,000) The African Start-Up Award – 3RD PRIZE – (€10,000)	The King Hassan II Great World Water Prize rewards excellence focused on "Cooperation and solidarity in the areas of management and development of water resources."[11] The African Start-Up Award was developed by the New York Forum Institute in partnership with France 24 and Le Point to support talent and the most creative initiatives, particularly those with a social impact component.
Mark Davies, founder and CEO, Esoko (Ghana): Esoko began in 2008, providing market prices over short message service (SMS) to smallholder farmers on development projects. Today, Esoko connects more than 1 million farmers to essential services – weather forecasts, agronomic advice, market linkages, and insurance coverage over a range of channels including SMS, voice SMS, and call center.	Mobex Innovation Africa (MIA) Award 2018 Impulse UM6P Start-Up Accelerator Blockchain Pitch Competition 2019 (USD$540.51) National ICT Initiatives Support Programme (NIISP) ($27,025.57)	Mobex Innovation Africa (MIA) Award for Governance and Social Good. IMPULSE is a 12-week acceleration program dedicated to innovative startups. IMPULSE aims to help entrepreneurs take their startups to the next level through equity-free cash prize of $250,000 to be shared between winning startups, investment opportunities, all-expenses-paid training in Morocco, access to business opportunities, resources, and mentoring. The Blockchain Pitch Competition is aimed at innovators with a passion for developing sustainable digital solutions leveraging technologies in blockchain, big data, AI, robotics & drones — to solve pressing challenges across their communities. NIISP is a government-funded program aiming to create a sustainable ecosystem for information and communication technologies (ICT) innovations and support ICT innovators and developers at large in Uganda. In 2019, 665 startups registered to receive funding; however, only 60 startups were selected to receive the grant.

[a] World Water Council. (2015) Nigerien social entrepreneur Abdou Maman wins the King Hassan II Great World Water Prize (US$ 100,000). Available at: https://www.worldwatercouncil.org/en/nigerien-social-entrepreneur-abdou-maman-wins-king-hassan-ii-great-world-water-prize-us-100000 (Accessed: 1 August 2020).

Esoko. (2018) Esoko wins Best Technology in Agriculture Award at Mobex Innovation Africa [Online]. Available at: https://esoko.com/esoko-wins-best-technology-agriculture-award-mobex-innovation-africa/ (Accessed1 August 2020).

Impulse. (n.d.) Home page [Online]. Available at: http://www.impulse.ma/ (Accessed: 1 August 2020).

Ministry of ICT & National Guidance (n.d.) 2019 award winners [Online]. Available at: http://niisp.ict.go.ug/2019-award (Accessed: 1 August 2020).

Africa Enterprise Challenge Fund. (n.d.) Competitions [Online]. Available at: https://www.aecfafrica.org/portfolio/competitions (Accessed: 1 August 2020).

Challenge funds: These funds are emerging as a critical financing source for entrepreneurs in the agriculture and food landscapes. They are often established by development partners and foundations and are typically announced and processed via funding windows with clear guidelines, timelines, and steps for applications. Usually, innovators have to demonstrate social impact and an ability to contribute a portion of the costs required for the scaling process. One example of a high-impact challenge fund on the African continent is the Africa Enterprise Challenge Fund (AECF), initially funded by the UK Department for International Development (DFID). Established in 2008, by 2019, AECF[6] has conducted 27 competitions, received over 10,000 applications, and approved funding to 266 business projects. This funding is matched by the businesses and catalyzes initiatives in the agriculture landscape. Its focus is on practical projects that are commercially viable and have a broad developmental impact on the rural poor.

Another challenge fund is Innovations Against Poverty, funded by the Swedish International Development Cooperation Agency (SIDA), which propels the private sector to develop products, services, and business models that can contribute to poverty reduction and combat climate change. In addition, the TradeMark East Africa Challenge Fund is focused on promoting cross-border trade in East Africa, ultimately to boost business and stimulate economic growth in the region.[8] Additional challenge funds include The AgriFI Kenya Challenge Fund for agrienterprises, the GIZ Innovation Challenge, the Access Agriculture Young Entrepreneur Challenge Fund, Famae Food Challenge, and the USAID Africa Trade and Investment Hub.

In-kind support: Beyond monetary contributions as donations and grants, companies can also benefit from a range of products and services offered at significant discounts or for free. These contributions include volunteer time, discounts on media advertising and bank fees, free airline tickets, books, and even office space. These donations should always be costed at market rates and actively tracked as an essential source of support for the organization.

AACE Foods benefited from in-kind support from the Business Innovation Facility (BIF), which operated in Nigeria between 2010 and 2019, managed by the Convention on Business Integrity. The facility engaged experts to provide market research, new product development, and branding support for the company at a time when it could not afford to engage external resources.

The growth phase

At this stage, the entrepreneur has proven that his/her business model is demand-driven and can generate significant revenues. There is also sustained customer interest and engagement, which leads to reduced perceived business risk from funders and financial partners. Investors are attracted by the company's ability to share financial statements that reveal positive cash flow and an enhanced ability to repay debt. At this stage, the company needs working and growth capital and can engage with private equity investors, banks, and other financial institutions.

- **Private equity:** Private equity (PE) is patient capital, which is structured as preferred debt and equity, invested primarily by institutional or accredited investors and family trusts. These investors typically expect to partner with the entrepreneur to restructure his/her business and request board seats to ensure corporate governance. They also expect to generate double-digit returns and to exit the investment after 5–7 years. According to *Critical Capital for African Agrifood SMEs*[12], there are more than 100 funds providing investment capital to agrifood small- to medium-size enterprises (SMEs) in Africa. The size of the investments can categorize funds (e.g., up to $1 million is small, up to $5 million is medium, and more than $5 million is large), by geographical reach (e.g., national, regional, global), or a distinction can be made according to the instrument/investment policy (e.g., wholesale agriculture, niche fund, venture capital, local/regional fund, fund of funds, etc.).

    Impact investors: This type of investor typically operates as venture capital or private equity. However, they explicitly require that your company engage in social change as part of its operations, either through its supply chain or distribution mechanism. They explicitly track indices such as employment creation, involvement of smallholder farmers (outgrowers), and women's empowerment to measure the intervention's impact, requiring evidence beyond a few success stories.

    There are many examples of successful PE investments in the African context. A few include:
    - In November 2018, Kobo360 raised $6 million from the International Finance Corporation.[13] In August 2019, the company

completed a Series A round, raising $30 million from local and international investors, including Goldman Sachs, and in 2020, it launched a $50 million equity round.
- In 2018, Cellulant raised $47.5 million[14] in its Series C round from TPG Growth's The Rise Fund to scale digital payments across Africa.
- In 2019, Twiga raised $23.75[15] million in a Series B equity round led by Goldman Sachs, with participation from existing investors, including the International Finance Corporation, TLcom Capital, and Creadev, and an additional $6 million in debt from OPIC and Alpha Mundi. Since its inception in 2014, the company has been able to raise a total of $55 million[16] in debt and equity.
- In early 2020, Kenya's agritech startup Farmshine raised $250,000[17] from US-based impact investor Gray Matters Capital's gender lens sector-agnostic portfolio – GMC coLABS. This funding will be used to hire and train personnel, including field agents, and to further develop the Farmshine platform, which connects farmers with the information, suppliers, and service providers they need to minimize costs and maximize harvests.

Box 6.1: THE FUND FOR AGRICULTURAL FINANCE IN NIGERIA (FAFIN)

Sahel Capital Agribusiness Managers Limited ("Sahel" or "SCAML") is a private investment firm focused on the food and agriculture sector in Nigeria to transform the agricultural finance landscape in Nigeria by investing in high-growth enterprises (SMEs) across the agricultural value chain in Nigeria.

In 2013, Sahel was selected to act as the fund manager for the Fund for Agricultural Finance in Nigeria (FAFIN), the sponsors of FAFIN include the Nigerian Federal Ministry of Agriculture and Rural Development, KfW Development Bank, the Nigeria Sovereign Investment Authority, the African Development Bank, CDC Group, and the Dutch Good Growth Fund as investors.

Since its inception, FAFIN has invested in seven companies across six states in Nigeria in the crop and livestock value chains, integrated processing operations, branded packaged foods, and related services. The private equity firm not only provides long-term, tailored finance to

> these companies, but also has a technical assistance facility for building management capacity, product certification, market research, ESG related initiatives, and developing smallholder farmer outgrower networks. FAFIN aims for a double bottom-line growth, achieving both commercial returns and significant developmental impact.
>
> FAFIN, through its portfolio companies, has created more than 900 direct jobs, with more than 120 roles for women and more than 2,000 jobs when factoring in seasonal and casual workers; directly impacted 9,000 smallholder farmers, with women making up more than 70% of these farmers; and indirectly impacted more than 65,000 individuals across a broad range of value chains through job creation, local crop sourcing, and engagement with the smallholder farmers and their beneficiaries.

Loans: Entrepreneurs in the growth phase may take loans to fund their working capital, typically structured as short- or long-term loans for significant projects, including constructing a new factory or purchasing major production equipment. Loans are often available from public entities, such as development banks and banks of agriculture at subsidized rates to encourage entrepreneurship in the food and agriculture sector. Loans are accessible through farmers' associations and cooperatives for small-scale farmers, but larger-scale businesses can often access these credit facilities directly. For example, the Global Innovation Fund provided a $2.5 million[18] loan to "to support the expansion of Babban Gona's innovative agriculture franchise model that seeks to sustainably improve the lives of smallholder farmers through comprehensive farming services."

Commercial banks across Africa also offer loans; however, interest rates are likely to be much higher, coupled with more extensive collateral requirements. You must weigh the options provided to you by commercial banks against the other funding opportunities that may be available to you to determine whether and when to engage with them.

Sometimes funders offer "repayable grants," which are effectively 0% interest loans. If they are denominated in a currency other than the country's currency in which the company operates (e.g., USD repayable grant in Nigeria), pay attention to currency risk during future repayment, given the costs associated with devaluations.

Maturity

At maturity, your business is well-entrenched as a respected player in the sector and can readily access debt capital; however, growth rates start to decline. Your ability to retain fresh funding and interest from funding sources is linked to your vision for the business, whether you are interested in exiting at this stage or still have the personal drive for the company.

Decline

If the company is struggling to innovate at this stage, it will eventually experience dwindling sales, die a slow death, or become a candidate for acquisition. Financial choices are critical even at this stage, including determining how much to spend on innovation versus reinvestment in or managing old assets. External financing is more difficult to obtain. However, sales of assets are a potential source of funds.

Table 6.3 Key Issues Assessed by Providers of Capital

Business Life Cycle	Key Issues for Providers of Capital
Startup	Entrepreneur's perceived integrity, grit, and sectoral knowledge ranks high. The attractiveness of the investment opportunity is also important. Sophisticated investors also want individuals who can pivot quickly and demonstrate agility as the reality of execution rips apart the intended business plan. Due diligence is primarily focused on background checks on the entrepreneur. The investor's perceived passion of the entrepreneur is important because this is a potential outward reflection of the resilience that the entrepreneur will need to carry through tough times, given that the investor only gets a return on investment if the entrepreneur is resilient.
Early Stage	Investor's assessment of initial execution track record and the entrepreneur's ability to onboard key team members and partners. The investor also assesses how quickly the entrepreneur is advancing towards "proving" the business model – as demonstrated by sales, key new customer accounts, or other appropriate metrics relevant for the particular business. Assessment of the team's ability to rapidly scale the business, while competing against substitute products/services is critically assessed.

(Continued)

Table 6.3 Continued

Business Life Cycle	Key Issues for Providers of Capital
Growth	Due diligence on governance structures becomes significantly more important – financial records and controls, statutory compliance, and completeness of appropriate legal documentation. Investors and/or lenders rely on these records to make decisions on valuation and/or creditworthiness of a company, as well as potential contingent liabilities that may exist. Perceived strengths and execution ability of the management team is important. Companies that have developed proper management teams are viewed as more favorable than those that are still founder-only led. Commingling of personal/company funds are red flags. Tax and statutory evasion could signal integrity issues – if companies do not honor statutory authorities' obligations, why would it honor commitments to unrelated investors? Due diligence also involves benchmarking performance against perceived peers – profitability, execution, governance structures, etc. – to determine the best companies to back. Lenders are very particular about the nature of collateral that would be accepted, and the cost of debt financing tends to be more expensive than that of mature companies.
Maturity	With growth companies, investors assume that governance structures may not be perfect, with mature companies, the default assumption is that these structures are in place. If the company is not well structured, this is a significant red flag. Due diligence (in addition to the issues assessed with growth companies) is conducted on areas in which incremental value creation can be achieved, including margin enhancements, business areas to close or other areas to focus on, cost reductions, possible geographic expansion, etc. A good benchmark of performance is the strength of operating cash flows. Lenders are also more comfortable because of likely robustness of assets for collateral and can be flexible on a range of related collateral options. Cost of debt financing also tends to be lower than that of smaller growth companies.
Decline or Reinvention	Similar to mature companies, substantive attention is paid to where incremental value creation can be achieved. Does the company also have under-levered product brands? Can a new management team rapidly drive through needed changes or reinvent the business? Is the company decline related to fixable organizational issues or fundamental changes in the industry dynamics that the company has not responded to effectively. Cost of debt financing also inches upward linked to perceived increased riskiness of the company and perhaps constrained or declining operating cash flow.

Steps for raising funding

While there are numerous funding sources for you to consider, the process of raising funding from these sources is challenging and demands a proactive fundraising strategy. Once you have determined your need for funds and appetite for the options available, there are four critical steps to achieve the desired results:

1. Ensure that your company is investment-ready.
2. Reach out to potential funders and engage, negotiate, and accept the funds, if appropriate.
3. Utilize the funds, track results, measure, monitor, report, and effectively manage relationships with funders.
4. Plan for exits and future funding.

Ensure that your company is investment-ready: External funders are only interested in engaging with organizations with strong credibility, governance structures, financial management systems, and controls and those that can demonstrate the ability to utilize the funds to achieve results. There are many reasons why investors decline applications from entrepreneurs in the food and agriculture landscape, as documented in *Critical Capital for African Agrifood SMEs*[19] report.

BOX 6:2: REASONS VC/PE FUNDS DECLINE APPLICATIONS FROM AGRIFOOD SMES[20]

- Insufficient growth perspective in the market, lack of proven product-market combination, business model not scalable
- Expected return (internal rate of return, [IRR]) too low, lack of ability to create shareholder value (e.g., 3–4x multiple on exit)
- Doubts about the entrepreneurs themselves (sometimes too old, not entrepreneurial, not technically qualified) or the management team in general. Often too dependent on one single person (key-man risk)
- Weaknesses beyond the firm's direct control, such as low quality and productivity of its suppliers (outgrowers), may jeopardize the business case
- Owner does not want to share control (let alone cede a majority stake)
- No shareholder structure or governance; may not even be legally established and registered

- Lack of viable exit strategy, in particular for the smaller ventures, or the owner/coowner objects to the fund (eventually) selling its stake to an outsider
- Financing needs too small, startups in particular
- Too little own investment by the entrepreneur ('skin in the game'). Sometimes the business is already heavily indebted or has a poor prior credit history
- Lack of proper accounting (let alone audited financial statements), management and reporting systems, hampering the due diligence and subsequent monitoring process
- Specific requirements relating to social impact or sustainable production are not met (e.g., no quantifiable social impact, use of ecologically unsustainable production methods)
- Financial, operational, tax, and legal due diligence reveals issues of noncompliance involving a (reputation or financial) risk
- Politically exposed persons among the company's owners or managers

As outlined in Chapters 2 and 3, proving a robust business model and instituting a strong board, composed of credible and capable directors, are critical prerequisites for obtaining funding. In addition, your company has to institute robust financial management systems, internal and external audits, boundaries and controls, and prepare financial reports demonstrating that it is financially healthy. Vital financial statements and controls that investors expect to see actively utilized within your company include the following.

- **Budgets:** Annual and quarterly budgets that enable financial planning and monitoring.
- **Balance sheets:** These outline the assets and liabilities of the entity at a particular date in a calendar year and enable the company to understand its financial position.
- **Profit and loss statements:** Enable the leadership to interpret the results of operations, sources of earnings, cost drivers, and profitability.
- **Cash flow statements:** Provide insights into the inflow and outflow of funds.
- **Financial snapshots:** Summarize key performance indicators based on business objectives.
- **Taxes:** Evidence of payment of all relevant taxes including VAT, payroll tax, corporate, state, and city taxes.

- **Audits:** Financial statements audited by a credible, independent accounting firm completed at least annually.

WORKSHEET 6.1: INVESTMENT READINESS

- Are you recognized as an innovative, credible, and accountable leader with an experienced and committed team?
- Do you have a positive track record based on what the team has done before or through this intervention?
- Do you have a proven market opportunity?
- Do you have a resilient business model?
- Do you have strong governance structures, including a committed board that the investor can work with?
- Can you demonstrate a healthy business: robust financial management systems and positive cash flow and the potential for an attractive upside for the impact investment?
- Can you demonstrate a feasible exit?
- What strategies do you have to mitigate the most significant environmental, social, and economic risks associated with your business model and operations?

Engage in the fundraising process: There is growing evidence that entrepreneurs in the food and agriculture landscape are raising funds both locally and internationally. However, the process requires patience, diligence, and commitment.

As an entrepreneur, it is important to understand the assessment process used by various capital providers to enable you to tailor your pitch and proposals appropriately. Providers of capital focus on different issues at different stages of a company's life cycle when assessing the attractiveness of providing equity or debt capital to a company. The assessment process for equity and debt providers of capital is different. Equity is focused on upside investment return potential, while debt providers are focused on downside protection. Funders who offer grants, challenge funds, and other investment instruments are more focused on the development impact story than commercial return possibilities.

Instead of outsourcing the fundraising process to a third-party, consider building the skills in-house by crafting your teaser and pitch deck, which lays out your vision for the company and financing needs. Grant applications are fairly straightforward but expect to complete several applications before successfully securing the most appropriate funding. In addition, expect an iterative process with the prospective funder, who will request changes to your recommendations, and a range of supporting documentation including your incorporation certificate, recommendations from your partners and other funders, audited financial statements and in some cases, letters from your bankers stating that your company is in good standing.

> **Box 6.3: RAISING FUNDS – THE REELFRUIT EXPERIENCE**
>
> Reelfruit, founded in 2012 by Affiong Williams, is a dried fruit processing, marketing, and distribution company in Nigeria. Its business model is focused on transforming otherwise wasted raw materials into well-branded snacks, thereby guaranteeing a market for farmers and increasing their incomes.
>
> In 2019, ReelFruit sold the equivalent of more than 400,000 pineapples that would have otherwise gone to waste. ReelFruit's product range comprises branded and retail-packed dried fruit and nuts snacks (pineapple, mango, fruit/nut mix, and cashews), sold at more than 350 outlets around Nigeria, through online channels, and to business-to-business market (hotels, airlines), answering to the growing demand for healthy snacks. The products are also exported to Europe (Switzerland and Belgium) and sold on Amazon.com.
>
> Like most entrepreneurs, Williams's first round of investment came from her savings as well as from friends and family. For the first year, she used her savings to fund the pilot of the business, validate the market, get customer feedback, and obtain all the required regulatory approvals to ensure that the company sell its products in supermarkets. After reaching these milestones, she then engaged her family to invest to help her rent space and hire the first set of staff.
>
> According to Williams, fundraising is an extremely challenging and draining process. There is typically a very low conversion rate so, it

requires a lot of spreading the net wide and talking and following up with a lot of people to convert. It is also extremely distracting and takes your attention from daily oversight of the business.

Three tips for entrepreneurs who are raising funds:

- *Be very diligent in keeping records. You need to demonstrate to investors that you are able to show how the business is growing and demonstrate what you have been able to do with the resources you have so far.*
- *Be very milestone-based in your funding requests. In my experience, it is easier and more manageable to ask for $100,000 from individuals or angels for milestone-based growth (for a set period of time), than $2 million from one large investor for large-scale expansion.*
- *Keep your investors up to date and engaged with your business. It is much easier to raise more money from an existing investor than a new one, so keeping your investors close to the business and sharing the highs and lows with them increases your chance of raising follow-on funding that will enable short- to medium-term growth.*

The African Enterprise Challenge Fund (AECF)[21] outlines a four-step process for engaging with entrepreneurs who require funding. This process is rigorous, and many entrepreneurs find it daunting. However, the process itself is a useful learning experience, and with each application, you will build skills and relationships that will be critical for your future growth.

- **Application:** The Initial Concept Note Assessment is the first stage in the submission of an initial application form in which the applicants write about their business ideas and their company(s) on a template provided. If shortlisted, the entrepreneur, is engaged in the preselected Concept Note Assessment stage. He/she will be asked to submit a more comprehensive business plan for assessment by the investment committee.
- **Selection:** During this stage, the AECF engages with and visits the shortlisted applicants preparing their business plans to understand their model and innovations better. The entrepreneur then submits the final business plans to be presented to an independent investment committee who meet and decide which company projects to fund.

- **Award:** This includes three components – an induction workshop, contracting, and disbursement. Successful businesses are invited to participate in an induction workshop, in which postaward processes, such as progress reporting, are discussed before contracting. The contracting phase is rigorous and formalizes obligations and deliverables committed to the grantee to meet all contracting terms and inform the disbursement of funds. AECF actualizes disbursements following a predefined schedule agreed at the time of contracting, which are dependent on key deliverables being met at each stage of the project.
- **Evaluation:** Grantees must self-report semiannually. They can, however, choose to report quarterly if they need funds disbursed. In addition, an AECF project manager leads monitoring and evaluation visits to the projects semiannually or quarterly to assess technical and financial progress.

To date, AECF has supported a range of food and agriculture businesses across Africa through its Agribusiness Africa Window (AAW) and its general windows. AACE Foods benefited from one of these windows, and the team worked through the rigorous four-step process outlined above.

For venture capital, private equity, or large loans, consider engaging a financial adviser and a lawyer if you do not have in-house capacity to support you through the process of valuing your company and negotiating from a position of strength.

It is important to recognize that some funders also offer technical assistance through which you can benefit from business planning and engage additional experts to support you through the valuation and negotiations phase. Also, actively engage your board members for insights and support during this process.

Consider the One Acre Fund experience in fundraising, from its inception in 2006 to date, as first outlined in *Social Innovation in Africa: A Practical Guide for Scaling Impact*.[22] Andrew Youn's fundraising journey, from life as an MBA student at Kellogg School of Management to an award-winning social entrepreneur, is very instructive. He and his team have successfully raised funds from more than 40 institutions for their work on the African continent. Below are a few of these examples.

Table 6.4 One Acre Fund's Fundraising Journey

Date	Funder	Amount
April 2006	Yale Business Plan Competition – Social Entrepreneurship Track	$50,000
May 2006	Echoing Green Scholarship	2-year stipend
2007	Draper Richard Kaplan Foundation	$300,000 (spread over 3 years)
2007	Mulago Social Investments	$100,000
2009–2016	Pershing Square Foundation	$6.5 million (directed into the permanent fund that provides essential capital for loans to farmers)
2010	Skoll Foundation	$750,000
2011–2015	Mastercard Foundation	$10 million
2013–2016	Bill & Melinda Gates Foundation	$11.6 million
2016–2019		$6 million
2012–2016	Barr Foundation	$4.7 million to support crop insurance and sustainable initiatives
2016	Children's Investment Fund Foundation	$12.6 million for nutrition interventions
2017	Global Innovation Fund	$15 million to drive smallholder farmer income improvements

Utilize the funds judiciously and effectively manage relationships with funders: Once the funding is received, it must be deployed in a transparent and accountable manner for its intended purposes. Managing money judiciously is just as difficult as raising it in the first place.

You have to monitor how the money is utilized actively, measure impact and results, and provide regular financial and impact reports to the investment or funding team. Investors often plan due diligence trips before and after their investments, and some expect board seats, depending on the size of their investments and the value that they believe that they can bring to the entity. For example, both Draper Richard Kaplan Foundation and Acumen require board seats as part of their investments into social innovations. In addition, Sahel Capital also actively engages on the boards of the companies in which it has invested. Impact investors also expect that the innovator will work diligently to achieve the preset targets to enable them effectively to exit within 3–7 years, depending on the prenegotiated terms.

According to Sahel Capital, Zebu Investments, and other fund managers engaged during the research for this book, the best investor experiences are linked to the vision, commitment, and drive of the founder and his/her team with cooperation from the board. Growth and profitability can result in successful exits and sizable profits for the family, team, and investors. All parties work diligently through a trusting relationship to achieve the aggressive targets, rooted on a foundation of integrity and shared goals.

Plan for exits and future funding: While there are few publicly available examples of successful fund exits on the African continent, anecdotal evidence suggests that these are occurring quietly. At the very least, entrepreneurs are repaying their loans or buying back their shares from private equity investors. A few companies have engaged in initial public offerings and are now traded on their countries' stock exchanges.

- In 2017, investors in the $100 million Agri-Vie Fund 1 exited from Fairfield Dairy, one of South Africa's most respected dairy companies. According to VentureBurn,[23] the Fund invested $4 million in Fairfield Dairy in 2009 and exited between 2.5 and 3 times the invested capital. According to Kevin Lang, founder and major shareholder of Fairfield Dairy, "Agri-Vie brought both risk and capital, as well as many other valuable contributions to the business, having actively worked with management to improve both revenues and profitability over the investment period." The fund has also exited a few other companies in the African food and agriculture including, Africa Juice in Ethiopia, Fairfield Dairy in South Africa, Hygrotech in South Africa, HIK Abalone Farm in South Africa, and Vida Oils in Mozambique and South Africa.

As clearly demonstrated by Phatisa[24], one of Africa's largest funds focused on agriculture and food, its commitment to social impact demands that it specifically tracks key indicators in its portfolio companies linked to the UN's Sustainable Development Goals (SDGs). Through monitoring the activities of its seven portfolio companies, one subsidiary fund and three exits completed, amounting to a footprint across 17 countries, it has listed the following highlights from investment to the end of Q4 2019:

- **SDG 1:** No poverty – raised and investing $367.5 million in food-related businesses in Africa through African Agriculture Fund and Phatisa Food Fund 2.

- **SDG 2:** Zero hunger – produced more than 3.5 million tons of food and food-related products in Africa.
- **SDG 5:** Gender equality – impacted more than 18,500 female employees and beneficiaries directly.
- **SDG 8:** Decent work and economic growth – impacted more than 85,000 smallholder farmers and micro, small, and medium enterprises linked to Phatisa food and food-related investment portfolio and associated technical assistance projects.
- **SDG 12:** Sustainable development – 100% of portfolio companies have environmental and waste management policies, and 63% have implemented recycling initiatives.

Summary

There are many financing options available to you and your teams for scaling your business. These include angels, venture capital, grants, accelerators, prizes, awards, fellowships, challenge funds, private equity, crowdfunding, and loans. However, you have to identify and assess these options and their suitability for your business at this time. You must also institute the financial systems and structures to demonstrate investment readiness. Once your company receives funding, it is imperative that you actively track and report on the impact of these resources and demonstrate a high level of integrity and accountability in the use of the funds.

Notes

1. Lagos Angel Network. *About us* [Online]. Available at: https://lagosangel-network.net/about-us/ (Accessed: 1 August 2020).
2. Mastercard Foundation. (n.d.) *Partnership aims to create 560,000 work opportunities for young entrepreneurs and smallholder farmers* [Online]. Available at: https://mastercardfdn.org/partnership-aims-to-560000-work-opportunities-for-young-entrepreneurs-and-smallholder-farmers/ (Accessed: 1 August 2020).
3. Bill & Melinda Gates Foundation. (2015) *How we work: Grant – Babban Gona Farmer Services Limited* [Online]. Available at: https://www.gatesfoundation.org/How-We-Work/Quick-Links/Grants-Database/Grants/2015/07/OPP1125909 (Accessed: 1 August 2020).
4. Blue Moon Ethiopia. (n.d.) *Home page* [Online]. Available at: http://www.bluemoonethiopia.com/ (Accessed: 1 August 2020).

5. Blue Moon Ethiopia. (n.d.) *Home page* [Online]. Available at: https://www.bluemoonethiopia.com/bluemoon-ventures/bluemoon-seed-fund/ (Accessed: 1 November 2020).
6. Co-Creation Hub. (n.d.) Home page [Online]. Available at: https://cchubnigeria.com/ (Accessed: 1 August 2020).
7. Growth Capital. (n.d.) *Home page* [Online]. Available at: https://gc.fund/ (Accessed: 1 August 2020).
8. Jackson, T. (2019) "9 startups selected for Village Capital's Agriculture Africa accelerator," *Disrupt Africa*, 6 Septemter [Online]. Available at: https://disrupt-africa.com/2019/09/10-startups-selected-for-village-capitals-agriculture-africa-accelerator/ (Accessed: 1 August 2020).
9. Village Capital. (2020) *Disrupt Africa: Ghanaian, Nigerian startups secure $50k funding after Village Capital agriculture accelerator* [Online]. Available at: https://newsandviews.vilcap.com/in-the-news/agriculture-africa-2019-peer-selected-startups-featured-in-disrupt-africa (Accessed: 1 August 2020).
10. Tomato Jos. (2014) *Kickstart Launch* [Online]. Available at: http://www.tomatojos.net/kickstarter-launch (Accessed: 1 August 2020).
11. World Water Council: King Hassan II Great World Water Prize. https://www.worldwatercouncil.org/en/king-hassan-ii-great-world-water-prize; (Accessed 11 November 2020)
12. van Manen, B. (2018) *Critical capital for African agri-food SMEs: A review of deman foor and supply of risk capital for agri-food SMEs in sub-Saharan Africa. Based on field studies in Kenya, Tanzania, Zamba and Mali*, p. 10 [Online]. Available at: https://www.icco-cooperation.org/en/wp-content/uploads/sites/2/2019/07/Critical-Capital-Web.pdf (Accessed: 1 August 2020).
13. Jackson, T. (2018) "Nigerian logistics startup Kobo360 raises $6m funding," *Disrupt Africa*, 7 December. Available at: https://disrupt-africa.com/2018/12/nigerian-logistics-startup-kobo360-raises-6m-funding/ (Accessed: 1 August 2020).
14. Cellulant Nigeria. (n.d.) Mission [Online]. Available at: https://www.cellulant.com/our-story/ (Accessed: 1 August 2020).
15. Miriri, D. (2019) "Kenya's Twiga Foods raises $30 million ahead of planned expansion," *Reuters*, 28 October [Online]. Accessed: https://twiga.ke/2019/10/28/twiga-foods-secures-30m-to-digitize-food-distribution/ Available at: https://www.reuters.com/article/us-kenya-twiga-idUSKBN1X71BL (Accessed: 11 August 2020).
16. Whitehouse David, *Africa Report*; Kenya: *Twiga Foods seeks funding to expand in East African market*; August 5th 2020. https://www.theafricareport.com/36366/kenya-twiga-foods-seeks-funding-to-expand-in-east-african

17 -market/#:~:text=The%20company%2C%20launched%20in%202014,to%20expand%20into%20West%20Africa. (Accessed 10 November 2020)
17 Mbabazi, E. (2019) "Kenya's Farmshine raises $250,000 funding," *Kenyan Wall Street*, 6 December [Online]. Available at: https://kenyanwallstreet.com/kenyas-farmshine-raises-250000-funding/ (Accessed: 1 August 2020).
18 Global innovation Fund. (n.d.) *Investments: – Babban Gona* [Online]. Available at: https://www.globalinnovation.fund/investments/babban-gona/ (Accessed: 1 August 2020).
19 van Manen, B. (2018) *Critical capital for African agri-food SMEs: A review of deman foor and supply of risk capital for agri-food SMEs in sub-Saharan Africa. Based on field studies in Kenya, Tanzania, Zamba and Mali*, p. 10 [Online]. Available at: https://www.icco-cooperation.org/en/wp-content/uploads/sites/2/2019/07/Critical-Capital-Web.pdf (Accessed: 1 August 2020).
20 van Manen, B. (2018) *Critical capital for African agri-food SMEs: A review of deman foor and supply of risk capital for agri-food SMEs in sub-Saharan Africa. Based on field studies in Kenya, Tanzania, Zamba and Mali*, p. 10 [Online]. Available at: https://www.icco-cooperation.org/en/wp-content/uploads/sites/2/2019/07/Critical-Capital-Web.pdf (Accessed: 1 August 2020).
21 Africa Enterprise Challenge Fund. (n.d.) *Competitions* [Online]. Available at: https://www.aecfafrica.org/portfolio/competitions (Accessed: 1 August 2020).
22 Nwuneli, N.Odidi,. (2016). Social Innovation in Africa: a practical guide for scaling impact. England: Routledge 2016.
23 Timm, S. (2017) "Agri-Vie Fund exit from Fairfield sees investors reap over three times in returns," *Venture Burn*, 3 July [Online]. Available at: https://ventureburn.com/2017/07/agri-vie-fund/ (Accessed: 12 July 2020).
24 Phatisa. 2020. Phatisa and partners acquire southern African agricultural solutions provider FES [Online]. Available at: https://www.phatisa.com/wp-content/uploads/2020/04/PHATISA-AND-PARTNERS-ACQUIRE-SOUTHERN-AFRICAN-AGRICULTURAL-SOLUTIONS-PROVIDER-FES-22.4.2020.pdf (Accessed: 12 July 2020).

7

SHAPING YOUR ECOSYSTEM

Introduction

No day is the same for the average entrepreneur in the food and agriculture landscape in Africa. Every day brings new adventures, challenges, opportunities, and risks. The entrepreneurs who can successfully ride this wave typically understand their ecosystem and have built stable support structures and enduring relationships.

This chapter highlights the complex ecosystems in which entrepreneurs in the food and agriculture landscape operate. It outlines opportunities and strategies for shaping policies, building strategic partnerships and coalitions, and navigating complicated relationships with national, state, and local governments as well as development partners, industry associations, and nonprofit organizations. Leveraging case studies, it provides practical frameworks that entrepreneurs can leverage to identify, build, and sustain valuable partnerships critical to scaling their businesses.

The entrepreneurial ecosystem

The entrepreneurial ecosystem[1], as defined by Prof. Colin Mason and Dr. Ross Brown is:

> a set of interconnected entrepreneurial actors (both potential and existing), entrepreneurial organizations (e.g., firms, venture capitalists, business angels, banks), institutions (universities, public sector agencies, financial bodies) and entrepreneurial processes (e.g., the business birth rate, numbers of high growth firms, levels of 'blockbuster entrepreneurship,' number of serial entrepreneurs, degree of sellout mentality within firms and levels of entrepreneurial ambition) which formally and informally coalesce to connect, mediate and govern the performance within the local entrepreneurial environment.

Given the high level of fragmentation in the African agriculture and food landscape, as outlined in Chapter 1, there are hundreds of actors in the ecosystem. Consider the example of the key players in the ecosystem for an average agribusiness, leveraging a model developed by Daniel Isenberg, formerly a professor at the Harvard Business School, and a global authority on entrepreneurship ecosystems.

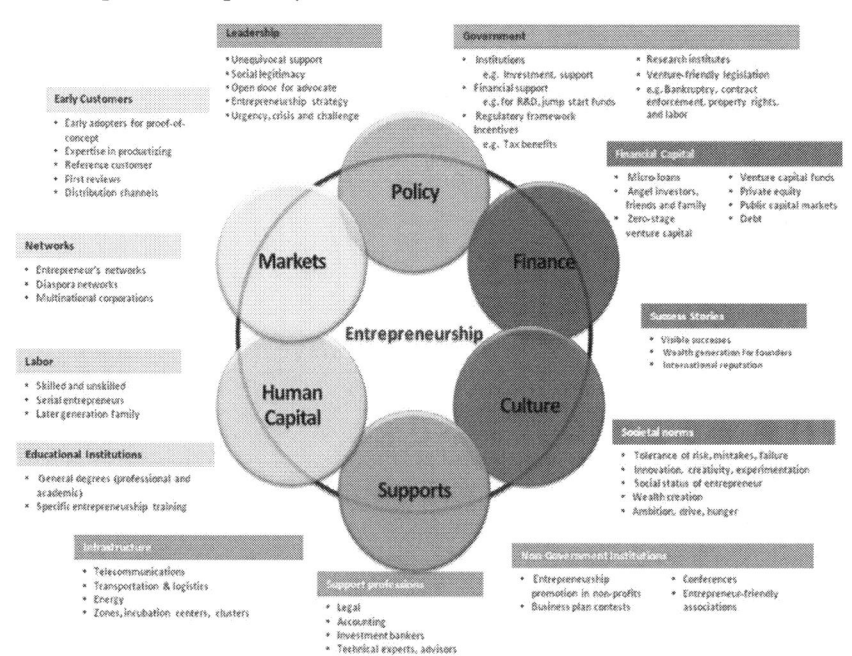

Figure 7.1 Domains of the entrepreneurs' ecosystem.

Mapping your ecosystem

Beyond identifying the players in your ecosystem, it is imperative that you further understand how their roles could affect your business in the short, medium, and long terms. A model created by Paul N. Bloom and J. Gregory Dees[2], which I utilized in my book, *Social Innovation in Africa*, serves as a sound basis for an assessment of key ecosystem stakeholders, with some refinement for the agribusiness and African context:

1. **Enablers/Drivers:** These are stakeholders who contribute directly to your momentum and success by providing financial, human, technical, and technology support.
2. **Beneficiaries:** These include the local host community, the input suppliers, the outgrowers, vendors and logistics providers who work with you, as well as customers and clients who purchase the products/services.
3. **Opponents/Critics:** These include groups or individuals support movements, initiatives, or products/services that directly frustrate your company's efforts, growth, and profitability in the short-, medium-, and long-term.
4. **Competitors:** These are other entrepreneurs or even bigger brands that serve the same markets, utilize the same resources, and/or source their raw materials from the same suppliers.
5. **Affected or influential bystanders:** These are individuals or organizations that have no direct influence or benefits from the enterprise in the short-term but, over time, could positively or negatively influence the success of the company or could be converted to any of the other categories above.

Depending on your community, focus value chain, and role within the value chain, your operating ecosystem could vary significantly by composition, concentration, and degree of activity of the stakeholders.

WORKSHEET 7.1: MAP YOUR ECOSYSTEM – INSERT SPECIFIC NAMES OF INDIVIDUALS OR INDIVIDUALS WHO FIT WITHIN EACH CATEGORY

Category	Enablers	Beneficiaries	Opponents	Competitors	Bystanders
Policy					
Finance					
Culture					
Supports					
Human – Capital					
Markets					

Identifying the key players in your ecosystem is the first step toward navigating the complexity of scaling your enterprise. The map will also provide critical insights into the potential collaboration opportunities for you to explore.

One of your key roles as a leader is to understand how best to navigate, engage, and thrive. You also need to periodically evaluate the landscape to understand the entry points, barriers, and levers required to scale your enterprise. It is also essential to recognize that your ecosystem will keep evolving. As a result, you will need to assess the dynamics every 6 months to understand what has changed and how to adjust, if required.

Engaging your ecosystem

One of the biggest misconceptions that entrepreneurs have is that they can build successful businesses without engaging with other entrepreneurs, community leaders, and the government. In reality, this is impractical in all sectors, especially in Africa's food and agriculture landscape. There are at least five critical lenses to consider.

The regulatory environment: The regulatory environment of any business ecosystem largely determines the success of the business. This notion was reinforced by Guarav Vijayvargiya of Seba Foods in Zambia, who stated that "policy is the most important part of sustaining a business."

Your enterprise is only as successful as the policies and regulations allow. According to Peris Bosire, the cofounder of FarmDrive, "if the government declares that only banks can lend money, FarmDrive will have to put a sizable amount of its resources towards paying for a banking license."

The COVID-19 pandemic reinforced the critical role that governments and policy environments play in either helping or hurting the growth and profitability of businesses on the African Continent. For entrepreneurs operating in countries such as Morocco, where the food and agriculture sector was identified as essential from the onset of the pandemic, there were minimal disruptions to the ecosystem. In countries with clear local sourcing policies, small- and medium-size enterprises (SMEs) even thrived during the lockdowns. Sadly, 50% of the SMEs on the Nourishingafrica.com hub, operating in countries with severe lockdowns, shut down temporarily, and many were concerned about their abilities to open again.

It is essential that you actively shape the political and regulatory environment in your community, country, and region. The following questions will help you gauge the political and regulatory environment in which you operate.

WORKSHEET 7.2: ASSESSING YOUR REGULATORY ENVIRONMENT

Policies
- Does the government prioritize agriculture and food?
- What value chains or agricultural issues are currently receiving political attention?
- What are the current and future policies that directly or indirectly influence your industry or entreprise?
- Are there intellectual property rights protection policies that protect innovations?
- Are the policies being implemented?
- Who are the key actors implementing the policies?
- Who are the champions and opponents of the existing and pipeline policies?
- Which regulatory agencies govern the food and agriculture landscape?

Taxes and Tariffs
- What are the current taxes levied on products and services and products in your country?
- Are there tax holidays for specific value chain products or activities?
- Are there exemptions from taxes for startups?
- Are there products that are banned from exports or imports?
- Are there associated export and import duties and quotas on certain products?
- What agencies do you need to engage for remittance of taxes and duties and for updates on changes in taxes and levies?

Lifecycle
- What is the lifespan of the current government at the state and federal level?
- What is the allowable compliance period for major policies and regulations?
- How will the end of the term of the current local, state, and federal governments affect your business?
- What is the validity period for tax exemptions and licenses obtained in your primary country of operations?
- What is the process for obtaining permits and licenses to sell products in the food and agriculture landscape?
- How frequently do they need to be renewed?

The competitive landscape: Assessing the competitive landscape is critical to understanding the existing and potential initiatives in the landscape that can undermine your commercial activities. According to Stiaan Wandrag of Tiger Brands, the competitive landscape is complex. It is not just between companies operating in South Africa but also includes retailers who import products from China and position them as private labels (with the brand names of the retailers), thereby bullying local manufacturers into lowering their prices.

According to a World Bank Report[3] on the conditions for market-based competition, more than 70% of African countries rank in the bottom half globally. Although there are at least 32 competition laws governing countries or regional blocks in Africa, few are vigorously implemented. Other nations that have no laws or processes are stalled at various stages of approval.

One exception is South Africa's Competition Act, which prohibits monopoly conduct, specifically if the company has at least a 45% market share. Its actions exclude other firms from the market and exercise excessive control over pricing. Violations could result in a penalty of 10% of the company's annual turnover from the previous financial year.

This law has implicated large companies in recent years, including Tiger Brands, Pioneer Foods, and Sime Darby. According to Tembinkosi Bonakele [4], head of South Africa's National Competition Commission, "food and agro-processing is an important focus area for the Competition Commission, and we are determined to root out exploitation of consumers by cartels."

It is imperative that you assess your competitive landscape every month, examining local and international players in your ecosystem.

WORKSHEET 7.3: ASSESSING YOUR COMPETITIVE LANDSCAPE

- Who are your competitors in the ecosystem?
- What is their core product/service offering?
- What are their competitive advantages compared to your product or service offerings?
- What are their limitations compared to your offerings?
- What unique advantages can your enterprise leverage to establish and sustain a leadership position in the market?
- Will there be an influx of new competitors in the medium- to long-term?
- What are the policies regarding competition in your country(ies) of operation?

Geography and infrastructure: Globally, entrepreneurs flourish in ecosystems that offer distinct advantages, including feeder roads, affordable energy, data services, innovation clusters, and other support services. FATE Foundation's 2019 entrepreneurship report, *From Start-Ups to Scale-Ups*,[5] identified infrastructural deficits and the location or locality of businesses as the third and fourth top inhibitors to the scaling of enterprises in Nigeria.

M-Pesa revolutionized the financial inclusion in Kenya and enabled the emergence of numerous agriculture and food businesses that have leveraged this platform. M-Pesa is an integral part of FarmDrive's business model, which it utilizes to provide financial products and services to smallholder farmers and agri-SMEs across Kenya.

Given that agricultural production activities are mostly concentrated in the rural areas on the African continent, entrepreneurs who are committed to leapfrogging must assess the landscape thoroughly and determine strategies for filling the infrastructure gaps.

WORKSHEET 7.4: ASSESSING THE GEOGRAPHY AND INFRASTRUCTURE STRUCTURES IN YOUR ECOSYSTEM

Telecommunications
- Who are the providers of digital, mobile, and broadband communication technologies/services?
- What is the cost of using the communications technologies for each of the providers identified above?
- What is the penetration of internet services in the region where you operate?
- How accessible and affordable are data services for your suppliers, distributors, and other stakeholders in your value chain?

Energy and Water
- What is the current electricity/energy situation in your operating environment?
- Are there alternative energy sources within your location?
- What is the cost of using alternative sources of energy?
- Are there water facilities needed to drive your business and that of your local suppliers?

Transport and Logistics
- What is the state of the feeder roads, airports, seaports, and railways that link your inputs to your factory and your product to markets?
- What is the cost of transporting your inputs, products/services within the country and other regions?
- Who are the major logistics providers in the industry?

Are there required storage facilities for your products?

Technology
- Are there new technologies that can drive the growth of your enterprise? Are they locally available?
- What is the cost of this technology?

What is the effect of leveraging the technology on your bottom line?

Professional networks and associations: Professional bodies and industry associations play a critical role in advocating for change in the business environment, disseminating industry information, and creating awareness for the products and services of its members. For example, the Consumer Goods Council of South Africa (CGCSA) is a nonprofit organization that engages MSMEs in the manufacture, retail, wholesale, and distribution of consumer goods. The CGCSA provides critical regulatory information that helps its members meet basic food hygiene and safety requirements and conducts advocacy projects in the sector. The activities of the CGCSA ensure that the entrepreneurs are retail-ready and help remove the sales and export barriers.

There are a growing number of associations by value chain, stage in the value chain, size of business, and focus of efforts – local, national, regional, international. There are also gender and youth-focused support groups emerging on the continent. As an entrepreneur, you will have to sift through the range of opportunities to engage and weigh the costs, benefits, and time commitments required.

> **Box 7.1: NOURISHINGAFRICA.COM – A DIGITAL HOME FOR ENTREPRENEURS IN THE AFRICAN FOOD AND AGRICULTURE LANDSCAPE**
>
> Nourishing Africa is a virtual home for agriculture and food entrepreneurs committed to attracting, empowering, equipping, connecting, and celebrating over 1 million dynamic and innovative entrepreneurs driving the profitable and sustainable growth of the African agriculture and food landscapes. Nourishingafrica.com serves as a platform for these stakeholders to scale their businesses, connect, and celebrate their successes

on the continent. The portal provides training support, access to funding and data, profiles of African cuisine and chefs, links to talent, and other resources to enable entrepreneurs to scale their businesses.

Some of the services that the hub provides to its members include:

- Knowledge platforms with access to training, data, and expert advice.
- Access and referrals to funders to scale their agriculture and food businesses.
- Marketplace to engage with potential customers, partners, and suppliers.
- Discounts on critical agricultural inputs, training programs, conferences, and services.
- Exclusive invitations to online and in-person training programs and access to members-only data and resources to build members' skills.
- Nominations for local and global speaking opportunities, media appearances, prizes, and awards.
- Free advertising and opportunities to showcase members' businesses.

In addition to these membership benefits, Nourishing Africa provides additional programs, tools, and resources for nonmembers, including "Ask an Expert," to fill knowledge gaps, "First Thursdays" to foster networking and community building, and "Business Resilience Toolkits."

The SUN Business Network is part of the Scaling Up Nutrition (SUN) Movement. It aims to engage and mobilize businesses at a global and national level to act and invest responsibly in improving nutrition. It provides a range of training programs, fairs, funding initiatives, and policy interventions to support businesses that produce nutritious food across Africa.

WORKSHEET 7.5: ASSESSING THE ASSOCIATIONS IN YOUR ECOSYSTEM

- Are there active professional associations in the landscape, by value chain, role in the food ecosystem, geographic focus, gender etc. whose activities align with the focus of your business?

- What are the significant activities and focus of the associations? Are there membership fees?
- What is the strength and composition of the association's membership?
- What are the association's role in the political and regulatory environment?
- What is their track record of integrity and their ability to influence policy?
- How relevant are the associations to supporting your vision?

Culture and norms of the society: As an entrepreneur building and scaling a business on the African continent, you must be prepared to contend with cultural, religious, and social norms. As reinforced by the story of AACE Foods in Box 7.2, traditional rulers and community groups play critical leadership roles in communities and serve as champions and key influencers. Similarly, input providers and aggregators typically have to "pay homage" to community or religious leaders before engaging networks of farmers; otherwise, they may meet with immovable roadblocks at every turn.

Be prepared to conduct a rigorous assessment of your host community to understand its leadership dynamics and sensitivities related to religion, gender roles, values and norms, prohibitions, and triggers. Understanding the dynamics in each community and navigating the issues that arise with wisdom, respect, and care are critical for operational success.

Box 7.2: CASE STUDY – AACE FOODS' NEW FACTORY

In 2012, AACE Foods purchased an old factory in a new location – the Sango Ota Area of Ogun State in Nigeria. A few days into the renovation process, some young men from the community shut down the process and blocked our operations. They claimed that we had not consulted them or obtained prior permission from the community leaders before commencing the renovations. Unsure of how to react, we immediately reached out to our experienced board members, who suggested that we engage the state government. Within 24 hours, we had engaged the commissioner for Industry Trade & Investment, who was extremely responsive and introduced us to the traditional rulers within the community. We immediately visited these traditional rulers to present our company and layout our plans to source from farmers and engage the youth as workers.

Shortly after the intervention, the young men disrupting our operations had dispersed, and they have not visited the factory ever since.

This early incident was very instructive in demonstrating the importance of engaging with critical stakeholders in the ecosystem. Since this incident, AACE Foods has actively participated in annual breakfast business meetings with the governor of Ogun State, maintained cordial relationships with community leaders, and joined industry associations, such as the Manufacturers Association of Nigeria and the Association of Food, Beverage and Tobacco Employers.

Forming strategic and sustainable partnerships

Ultimately, the purpose of mapping out key players in the ecosystem and assessing the entrepreneurial landscape is to establish strategic and sustainable partnerships with key actors in the ecosystem. Armed with the map of industry players and a critical understanding of the landscape, you must work diligently to establish relevant collaborations within and outside the sector. Indeed, this is the only way by which you build your networks, gain more recognition, and shape the ecosystem.

A critical challenge that you need to overcome when attempting to establish partnerships is the significant distrust among actors. This is particularly critical in the agriculture sector, in which the stakeholders are largely fragmented and naturally work in silos. To counter the high levels of distrust and skepticism, invest in building a strong brand name based on integrity in your local community. This will ensure that any background check that a potential collaborator is likely to conduct proves positive.

Determine how to engage with governments at the local, state, federal, and even regional level in some capacity. Governments can serve as your customers in the business-to-government models, addressed in Chapter 2, and provide financial and in-kind support. In addition, they can provide an enabling environment for your business to scale, including introducing appropriate policies and regulations, and providing financing through development banks or grant programs. Similarly, they can use their convening power to attract other partners into a value chain and foster partnerships.

If you are engaged in the provision of inputs such as seeds, fertilizer, and mechanization services, you will sometimes work closely with governments to pilot and scale your initiatives, either as suppliers for subsidy programs or partners, leveraging government extension support. When the government is stable and credible, all parties are happy, and you may be branded as an insider; however, you run the risk of getting sidelined and even destroyed if the government changes. Too often, entrepreneurs are punished by the incoming administration for their strong alliances with the previous government. As a result, you must navigate your relationships with governments carefully and stay neutral during elections.

> **Box 7.3: CASE STUDY: AFRICA IMPROVED FOODS RWANDA**
>
> Africa Improved Foods (AIF) is a joint venture and a public-private partnership aimed at tackling malnutrition in East Africa by producing nutritious, high-quality foods for those who need it most. Established as a joint venture between the government of Rwanda and a consortium of Royal DSM; FMO, the Dutch development bank; DFID Impact Acceleration Facility managed by CDC Group and International Finance Corporation (IFC); and the private sector arm of the World Bank Group, AIF is focused on reducing malnutrition in Africa.
>
> Since its establishment in 2017, with an investment of $60 million from the consortium partners, AIF had exceeded supply targets for the World Food Programme and the government of Rwanda. It has also launched five new commercial products in three countries and is working with 35,000 smallholder farmers in Rwanda.
>
> AIF recognized that to solve malnutrition and stunting in Rwanda, it needed to work closely with the Ministry of Health, the Ministry of Agriculture, etc. According to its CEO, Amar Ali, "having the government as the company shareholder ties the knot and works for both of us ... When it comes to the actual running of the business, they don't get involved. That's how they would like it, and that's how we would like it."

> AIF is shaping the ecosystem through partnerships. It is setting strict quality standards (<5ppb aflatoxin) for incoming maize and providing postharvest support to smallholder farmers, which has led farmers and other maize buyers to improve their quality management. It also expects that its postharvest work with farmers in the coming years will reduce total aflatoxin rejections to roughly 10%.

Key ecosystem changes required in the African agriculture and food landscape

Building on the constraints and promising trends shared in Chapter 1, it is imperative that entrepreneurs work with other stakeholders in the public, private, nonprofit, faith-based, and development communities to address the challenges and maximize the opportunities in this ecosystem. While all of the issues that require cross-sector stakeholder collaborations and intervention cannot be fully addressed in detail in this book, a few burning issues that urgently need attention and action are identified below.

Ensuring demand-driven research – Revamping African research institutions: In most countries in the region, agricultural research and development (R&D) remain divorced from the farmers, processors, and consumers who ultimately determine the relevance, level of adoption, and impact of that research. Unfortunately, instead of being guided by the needs of the market, research by the region's more than 60 dedicated agricultural research institutions in West Africa alone has been influenced by researchers themselves and donor interests. This limits its relevance and curtails data-driven policymaking for agriculture, which relies on a solid base of credible research.

There is an urgent need to bridge this gap between the market, research, and policy. To change course, West African countries could benefit from the experiences of countries that have become net exporters of food such as Australia and Brazil.

In Australia, research priorities are driven by farmers and food processors who themselves help fund that research. For example, the Australian Export Grains Innovation Centre (AEGIC)[6] is primarily funded by farmers who benefit from the improved seed varieties, market intelligence, and market linkages that are generated. More specifically, Australian wheat farmers willingly pay a levy at the point of exporting improved wheat varieties developed by the AEGIC to serve the particular quality requirements of international customers, including global noodle manufacturers.

Similarly, AgriBio[7], a joint initiative of the Australian government, through the Department of Environment and Primary Industries, and La Trobe University is structured as a private-public partnership. It devotes 70% of its activities to address challenges posed by the private sector. In comparison, 30% is focused on regional priorities in plant, animal, and microbial biosciences as well as biosecurity research and diagnostics.

West African research institutions should consider AgriBio and AEGIC's approaches to private-sector engagement. By introducing export-focused levies linked to innovative research, establishing private-sector advisory boards, private-public partnerships, and challenge funds to drive demand-driven research, they can diversify their funding bases from limited government budgets or aid agencies with set agendas and ensure widespread impact.

Cote D'Ivoire is already moving in this direction. Private producers predominantly fund its National Centre for Agricultural Research (CNRA) through the Interprofessional Fund for Agricultural Research and Advice (FIRCA)[8], which allocates at least 75% of the membership fees paid by producers of commodity crops to research serving that commodity. The remaining funds are allocated to a solidarity fund to serve sectors (mostly food crops) unable to raise sufficient funding through their own membership fees.

Second, West African research institutions need to be restructured to ensure that they embrace technology and performance-driven cultures. The Brazilian Agricultural Research Corporation (Embrapa) demonstrates the impact that research institutions can make in a country. Since its inception in 1973, Embrapa[9] has played a pivotal role in transforming the formerly degraded Cerrado lands into a region that now accounts for nearly half of Brazil's grain production. Its innovations in livestock have quadrupled the beef and pork supply and increased the chicken supply twenty-twofold, making Brazil one of the largest animal protein exporters globally.

Embrapa's mandate in 2018, "a technological innovation enterprise focused on generating knowledge and technology for Brazilian agriculture,"[10] reveals an organization that has continued to reinvent itself to remain dynamic and relevant to the needs of food producers locally and globally.

This same ethos must drive African agricultural research if it is to meet the unique needs of farmers and agribusiness stakeholders operating across the region and address the severe challenges that they face today and in the future.

Changing entrenched systems with interests protecting the status quo will not be easy. However, political will and a shared vision from governments, development partners, the private sector will ensure this catalytic ecosystem change.

Addressing malnutrition and promoting healthy diets: Research from the Institute for Health Monitoring and Metrics (IHME)[11] at the University of Washington links suboptimal diets low in vegetables, nuts, and seafood, or high in processed meats and sugary drinks to a higher risk of disease and disability. Essentially, these diets cause more deaths than any other risk factors in the world, including tobacco smoking.

There are a range of public, private, and nonprofit interventions on the African continent to promote healthier diets and address malnutrition being led by organizations such as the Global Alliance for Improved Nutrition (GAIN), the SUN Business Network, the Rockefeller Foundation, the Bill & Melinda Gates Foundation, and Sahel Consulting. However, changes in consumer behavior are not happening fast enough, reinforced by the high rates of stunting for children under the age of 5, anemia in women, and rising rates of noncommunicable diseases, such as diabetes and high blood pressure among the adult population.

Two of the ten critical transitions identified by the Food and Land Use Coalition (FOLU)[12] in Figure 7.2 are rooted in healthy diets characterized by a predominately plant-based diet that includes some "protective foods," such as fruits, vegetables, whole grains, and a diversified protein supply as well as reduced consumption of sugar, salt, and highly processed foods. A diversified protein supply includes aquatic, plant-based, insect-based, and laboratory cultured sources. These perspectives are echoed by Prof. Walter Willett[13] of the Harvard T.H. Chan School of Public Health, who states that

SHAPING YOUR ECOSYSTEM

FINANCIALS KEY
🏆 Economic prize by 2030   💰 Annual additional investment requirements to 2030   📊 Business opportunity by 2030

Ten Critical Transitions		Essential Actions	Financials (by 2030)
Healthy Diets	Global diets need to converge towards local variations of the "human and planetary health diet" – a predominantly plant-based diet which includes more protective foods (fruits, vegetables and whole grains), a diverse protein supply, and reduced consumption of sugar, salt and highly processed foods. As a result, consumers will enjoy a broader range of high-quality, nutritious and affordable foods.	**Government**: Establish and promote planetary and human health dietary standards through repurposed agricultural subsidies, targeted public food procurement, taxes and regulations on unhealthy food. **Business**: Redesign product portfolios based on the human and planetary health diet	🏆 $1.28 trillion 💰 $30 billion 📊 $2 trillion
Productive & Regenerative Agriculture	Agricultural systems that are both productive and regenerative will combine traditional techniques, such as crop rotation, controlled livestock grazing systems and agroforestry, with advanced precision farming technologies which support more judicious use of inputs including land, water and synthetic and bio-based fertilisers and pesticides.	**Government & Business**: Scale up payments for ecosystem services (soil carbon/health and agrobiodiversity) plus improve extension services (training and access to technology, seeds, etc.) **Business & Investors**: Shift procurement from buying commodities, to investing in sustainable supply chains; deploy innovative finance to reach currently underfinanced parts of supply chains	🏆 $1.17 trillion 💰 $35–40 billion 📊 $530 billion
Protecting & Restoring Nature	Nature must be protected and restored. This requires an end to the conversion of forests and other natural ecosystems and massive investment in restoration at scale; approximately 300 million hectares of tropical forests need to be put into restoration by 2030.	**Government**: Put in place and enforce a moratorium on the conversion of natural ecosystems, and give legal rights and recognition to the territories of indigenous peoples **Government**: Scale REDD+ to $50 billion per year by 2030 if results delivered and establish a Global Alliance Against Environmental Crime **Business**: Establish transparent and deforestation-free supply chains and demand the same of suppliers	🏆 $895 billion 💰 $45–65 billion 📊 $200 billion
A Healthy & Productive Ocean	Sustainable fishing and aquaculture can deliver increased supply of ocean proteins, reducing demand for land and supporting healthier, and more diverse, diets. This is only possible if essential habitats – estuaries, wetlands, mangrove forests and coral reefs – are protected and restored and if nutrient and plastic pollution are curbed.	**Government**: Protect breeding grounds, end both illegal fishing and overfishing, and provide tribe/ access rights to artisanal fishers **Government & Investors**: Develop new approaches and business models for insurance against catastrophic events affecting fisheries (storms, warming events, pH collapse) and for compensating poor fishermen for the cost of fish stock recovery	🏆 $350 billion 💰 $10 billion 📊 $345 billion
Diversifying Protein Supply	Rapid development of diversified sources of protein would complement the global transition to healthy diets. Diversification of human protein supply falls into four main categories: aquatic, plant-based, insect-based and laboratory-cultured. These last three sources alone could account for up to 10 percent of the global protein market by 2030 and are expected to scale rapidly.	**Government**: Use public procurement to secure long-term offtake for alternative protein sources **Government**: Increase R&D spending in alternative proteins (especially those with large benefits for lower-income consumers) and ensure that the resulting intellectual property remains in the public domain	🏆 $240 billion 💰 $15–25 billion 📊 $240 billion
Reducing Food Loss & Waste	Approximately one third of food produced is lost or wasted. To produce this food that is never eaten by people requires an agricultural area almost the size of the United States. Reducing food loss and waste by just 25 percent would therefore lead to significant benefits relating to environmental, health, inclusion and food security.	**Government**: Regulate and incentivise companies to report on and reduce food loss and waste **Investors**: Finance income-sensitive, climate-smart storage technologies	🏆 $455 billion 💰 $30 billion 📊 $255 billion
Local Loops & Linkages	With 80 percent of food projected to be consumed in cities by 2050, what urban dwellers choose to eat and how their needs are supplied will largely shape food and land use systems. This transition sets out the opportunity to strengthen and scale efficient and sustainable local food economies in towns and cities.	**Investors**: Invest in emerging technologies and innovations which will close the food system loop **Government**: City governments to foster local circular food economy through targeted public procurement and zoning	🏆 $240 billion 💰 $10 billion 📊 $215 billion
Harnessing the Digital Revolution	Digitisation of food and land use systems is occurring through gene editing techniques, precision farming, and logistics and digital marketing tools, enabling producers and consumers to make better, more informed choices, and to connect to the value chain rapidly and efficiently.	**Government**: Open access to public sector data (e.g. on national land registries, fisheries, agriculture, soil health etc.) and regulate and incentivise the private sector to provide open source data where appropriate **Civil Society**: Create, maintain and communicate results from real-time platforms for transparency, as is currently done through Global Forest Watch	🏆 $540 billion 💰 $15 billion 📊 $240 billion
Stronger Rural Livelihoods	Underlying all ten critical transitions is a vision of rural areas transformed into places of hope and opportunity, where thriving communities can adapt to new challenges, protect and regenerate natural capital and invest in a better future. Ensuring a just transition.	**All**: Establish public-private-philanthropic partnerships to train a new generation of young farmer entrepreneurs over the next decade **All**: Scale up rural roads and digital investments to drive productivity, end rural isolation and, in particular, initiate a global campaign for renewable electricity access for all **Government**: Safety nets for individuals and stranded communities to ensure a just transition	🏆 $300 billion 💰 $95–110 billion 📊 $440 billion
Gender & Demography	Women can be enormously powerful in shaping food and land use systems, thanks to their central role in agriculture and in decisions concerning nutrition, health and family planning. Making sure women have equal access to resources, such as land, labour, water, credit and other services, should be central to policies concerning the ten critical transitions, including by accelerating the demographic transition to a replacement rate of fertility in all countries.	**All**: Invest in maternal and child health and nutrition as well as education for women and girls **All**: Ensure access to reproductive health services and products	🏆 $195 billion 💰 $15 billion 📊 n/a

Figure 7.2 Growing better: Ten critical transitions to transform food and land use.

transformation to healthy diets by 2050 will require substantial dietary shifts. Global consumption of fruits, vegetables, nuts and legumes will have to double, and consumption of foods such as red meat and sugar will have to be reduced by more than 50%. A diet rich in plant-based foods and with fewer animal source foods confers both improved health and environmental benefits.

While promising at face value, the challenge is to ensure the affordability and availability of nutritious food, especially for low-income populations and the most vulnerable. This will require collaborative action from the private, public, and nonprofit sectors. For example, as cited by the FOLU report, African private-sector companies will have to reformulate their processed-food offerings to minimize unhealthy ingredients, including sugar. They will have to invest in producing healthier alternatives while leveraging locally sourced produce, including legumes and vegetables.

Governments will have to create incentives for companies and consumers to change behavior, leveraging taxes as incentives or disincentives. They will also have increase research and development spending on alternative protein sources for low-income populations, ensuring the broad-based commercialization of these efforts. In addition, they will need to invest in public infrastructure, including energy, storage, road, and cold chain networks, to enable private-sector efforts to minimize the high rates of postharvest losses of fruits and vegetables and minimize nutrient loss.

Nonprofits, faith-based organizations, the media, and civil society will have to hold the government and private sector accountable to ensure transparent and responsible leadership. In addition, they can play a critical role in raising consumer awareness and empowering communities to make more informed food choices.

Raising standards and eliminating food fraud: According to the United States' Grocery Manufacturers Association[14], food fraud affects approximately 10% of all commercially sold food products and costs the global food industry between $10 billion and $15 billion annually.

Data focused on the African continent is not as readily available, but what exists is alarming. A 2018 study by the Confederation of Tanzania

Industries[15] estimates that more than 50% of all goods, including food, drugs, and construction materials, imported into Tanzania are fake. Anecdotal evidence suggests that rates could be between 10% and 50% depending on the food category and the country.

As an agroprocessor and cofounder of Nigeria's AACE Foods and Sahel Consulting, I have observed first-hand the magnitude of the food fraud crises and how supermarket shelves and open-air market stalls are too often stocked with counterfeit products. In Nigeria, there is milk powder with no animal protein. In Kenya, there is vegetable oil made of recycled oil unfit for human consumption. In Ghana, the palm oil is laced with a food coloring called Sudan IV that is widely recognized as a carcinogen. In Uganda, formalin – an embalming agent – is used to keep meat and fish free from flies and seemingly fresh for days. Across Africa, there are incidences of plastic rice or nothing more than discarded rice chaff, packaged as high-grade rice, and corn powder dyed with Sudan IV, labeled as chili pepper.

Given that most of the counterfeit products in Africa are staples consumed to fulfill daily dietary needs, they are likely contributing to the rising levels of malnutrition and cancer on the continent. When parents living on just dollars a day believe they are buying their children milk, and that milk has no animal protein, the impacts on child development can be devastating. Indeed, there is no way to know to what extent food fraud is contributing to stunting, which affects 34%[16] of African children under the age of 5, with life-long impacts on physical and intellectual development.

Several issues are driving the rise of food fraud in Africa. First, the increasing complexity of food systems, ingredients with long supply chains, and varying levels of scrutiny and standards makes it exceedingly difficult to trace the origins of food products. Second, local manufacturers face increased competition from cheaper imports, which often have lower standards for African destinations, and so they may use inferior or even unregulated ingredients in their products to reduce their production costs. Third, weak regulatory standards, systems, and tracking mechanisms create loopholes in which counterfeiters can thrive.

We do not yet know the full economic or health impacts of food fraud. Still, it seems inevitable that, if left unchecked, this trend will exacerbate the nutrition and health challenges facing many countries in Africa. It could derail efforts to build healthy and vibrant local food systems.

Fortunately, some African countries and companies are beginning to take action. Many have signed on to CODEX[17] – a World Health Organization/Food and Agriculture Organization initiative that sets global standards for safe food. South Africa has enforced[18] extensive labeling regulations. It has prohibited Sudan I–IV as a food coloring, and regulatory agencies have removed products with this dangerous chemical from shelves. The Kenya Bureau of Standards and the Ghana Food and Drug Authority are likewise raising awareness[19] among traders and retailers about the harmful effects of food fraud.

But much more needs to happen. The cost of prevention is much lower than that of inaction, and everyone has a critical role to play. Indeed, only by exercising our collective political will and commitment to the health, food security, and economic development of Africa can we halt food fraud and rebuild the public's trust in food.

- African governments must set high regulatory standards for food content and labeling and track and prevent counterfeitly imported and locally produced food. Like the global war waged against counterfeit drugs, actions against food fraudsters must be bold, swift, and unrelenting.
- The African Union should establish a Food Fraud Network, similar to that of the European Union, to detect cross-border fraud and train food inspectors, police, customs officers, and others. The African Regional Economic Communities must sign formal trade agreements with their counterparts in Europe, Latin America, Asia, and the Americas focused on food safety and food fraud standards.
- The media must create broad-based awareness among citizens and empower them to identify counterfeit foods.
- Consumers must remain vigilant, raise alarms once food fraud is detected, and demand better protection by regulatory agencies and the government.
- Food manufacturers and agribusinesses, both local and global, must institute and enforce high standards of integrity and accountability. In addition, industry associations must become watchdogs and invest in self-regulation and self-policing to curb bad behavior.

Transforming agribusiness education: The talent gap in the African agriculture and food landscapes is a critical issue that continues to limit the growth of agribusinesses. The outdated curriculum being utilized by agricultural universities, the focus on teaching agriculture as science courses instead of business courses with limited use of technology and innovation, and the lack of practical applications and engagement in internships limits youth participation. For example, only 4.34% of graduates from Nigerian universities study agriculture or agri-related courses, and most of these graduates admit that this was their third or fourth choice.

In addition, the agricultural extension and information delivery system is weak, with a limited knowledge transfer from extension officers to farmers, resulting in a farming workforce that is largely underskilled and struggles to adopt cutting-edge technology. There is an urgent need for the private sector to partner with the Ministries of Education and the universities commissions in our different countries to update the curricula used in agricultural universities to include recent trends, management skills, and practical work experiences in agribusinesses. In addition, private organizations and institutions should support talent development efforts by providing internship and mentorship opportunities for undergraduate students. The government should also create national capacity-building programs for policymakers to empower them with the knowledge and skills to design sustainable policies for talent development in the agriculture and food ecosystem.

Repositioning African food globally: Sparse survey results on global preferences for food, which often excludes food from African countries or gives the entire region the lowest ratings, categorized food from Africa as one type of food instead of the diversity of 54 countries. Sadly, this global perception is seeping into the continent as well, with young Africans opting for fast food from international chains such as KFC, Dominos, and even Krispy Cream, making significant inroads into major African cities.

This means the world is missing out on extraordinarily diverse and tasty cuisine from across Africa, which is highly nutritious. More importantly, it means opportunities to foster greater global awareness about the culture and rich heritage of countries in Africa and promote cross-cultural learning are being limited.

These various findings, coupled with my investment in the transformation of the African agricultural and food landscapes over more than two decades, have led me down an unusual research path to discover how to change the disparity in the global appreciation and acceptance of African food.

Primarily, it will require African entrepreneurs to invest in branding and storytelling, innovation, and partnerships to accelerate broad-based awareness of African food. In addition, African governments must create an enabling environment and foster linkages with diaspora populations to attract the support required to position our food and change mindsets.

The rise of Japanese cuisine on the global stage presents some very compelling lessons for committed stakeholders in key African countries. As is the case with Ethiopian, Senegalese, or Nigerian cuisine today, in the 1960s and 1970s, awareness and appreciation for Japanese food were limited overseas. Then, the Japanese government acted because they recognized the critical role that food played, not only in creating an appreciation for a country and its people and building cultural bridges, but also in generating employment and demand for food from the country, they acted.

In 2003, the Japanese government expanded the scope of the Japanese External Trade Organization (JETRO)[20] to include branding, certification of restaurants and ingredients, and public relations. In addition, through strategic and deliberate interventions by the government, traditional Japanese cuisine was added to the UNESCO Intangible Cultural Heritage List in 2012. The Japanese Ministry of Health, Labour, and Welfare also produced studies showing the health benefits of Japanese food and the links between the cuisine and its life expectancy rates, which, at 85.7 years, is the second highest in the world.

The results of these efforts are very compelling. In 2006[21], there were approximately 20,000 Japanese restaurants outside Japan. By 2017, this number had increased five-fold to close to 120,000. Beyond the growth in restaurants, today, most leading supermarkets in the United States and the United Kingdom sell sushi. In addition, Japanese food attracts the highest ratings in foreign food preferences. This rapid rise of Japanese food in the global food landscape demonstrates what is possible, through strategic and coordinated efforts from critical stakeholders in the private and public sectors and what can be replicated by Cote d'Ivoire, Ethiopia, Nigeria, and Senegal, which have unique cuisines across the different ethnic groups and regions with distinct flavors.

Ethiopian cuisine already has a head-start, as there are more than 350[22] restaurants in the United States, driven by the immigration trends in the 1980s. Like Japanese food, Ethiopian food is considered extremely healthy. The injera made of teff is gluten-free and high in fiber, calcium, iron, and protein, making it very appealing to the growing global vegetarian and vegan community. This cuisine has the potential to gain widespread acceptance.

The momentum for food from other African countries is also building, with the emergence of celebrity chefs opening high-end restaurants in global cities. For instance, Senegalese chef Pierre Thiam, through his restaurant Teranga in New York City and his company Yolélé Foods, is committed to sharing African culture through food and promoting fonio as an attractive substitute to quinoa. Kwame Onwuachi, a second-generation Nigerian-American chef based in Washington DC, was named on the "TIME 100 List." Now is the time to build on this momentum and ensure that varied, healthy, and tasty African dishes are promoted globally.

Indeed, repositioning African food globally will not only increase the demand for food sourced from African farmers and create jobs for Africans locally and internationally, but it will also build more bridges between Africa and the rest of the world, breaking stereotypes and biases and, ultimately, enabling a deeper appreciation of our connections as humans.

Summary

Your interactions and involvement in the ecosystem in which you operate largely determine your access to critical resources, including inputs, markets, talent, funding, technology, mentorship, capacity development, data, and the knowledge needed to scale your business. Strategic partnerships expose you to new business opportunities and perspectives, provide new knowledge and technical expertise, and ensure long-term resilience and sustainability. In addition, your ability to shape the policy environment and government responses to the sector, especially the need to ensure an enabling environment for businesses like yours to thrive, is also heavily hinged on your active engagement in the ecosystem.

As you navigate this complex ecosystem, recognize the importance of building partnerships with government agencies, competitors, development partners, research and academic institutions, and industry associations. However, be extremely strategic and disciplined to ensure that this engagement does not detract from the leadership of your own company.

Notes

1 Mason, C. & Brown, R. (2014) *Entrepreneurial ecosystems and growth oriented entrepreneurship* [Online]. Available at: http://www.oecd.org/cfe/leed/Entrepreneurial-ecosystems.pdf (Accessed: 11 August 2020).
2 Bloom, P.N. & Dees, J.G. (2008) "Cultivate your ecosystem," *Stanford Social Innovation Review* [Online]. Available at: https://ssir.org/articles/entry/cultivate_your_ecosystem (Accessed: 11 August 2020).
3 World Bank Group. (2016) *Breaking down barriers: Unlocking Africa's potential through vigorous competition policy* [Online]. Available at: http://documents1.worldbank.org/curated/en/243171467232051787/pdf/106717-REVISED-PUBLIC-WBG-ACF-Report-Printers-Version-21092016.pdf (Accessed: 11 August 2020).
4 Reuters. (2017) "South Africa's antitrust watchdog seeks fine for Unilever," *Reuters*, 1 March [Online]. Available at: https://www.reuters.com/article/us-unilvr-pricefixing-safrica-idUSKBN1684U0?type=companyNews (Accessed: 11 August 2020).
5 FATE Foundation, From Start-Ups to Scale-Ups: A Review of Business Scale-up Activities in Nigeria" https://www.fatefoundation.org/download/from-startups-to-scaleups/. (Accessed July 20th 2020).
6 Australian Export Grains Innovation Centre. (n.d.) *Home page* [Online]. Available at: http://aegic.org.au/ (Accessed: 11 August 2020).
7 AgriBio Centre for AgriBioscience. (n.d.) *Home page* [Online]. Available at: https://www.agribio.com.au/ (Accessed: 11 August 2020).
8 ASTI. (2017) *Agricultural R&D indicators factsheet: Côte d'Ivoire* [Online]. Available at: https://www.asti.cgiar.org/sites/default/files/pdf/CotedIvoire-Factsheet-2017.pdf (Accessed: 11 August 2020).
9 Embrapa. (n.d.) *About us* [Online]. Available at: https://www.embrapa.br/en/web/portal/about-us (Accessed: 11 August 2020).

10 Embrapa. (n.d.) *About us* [Online]. Available at: https://www.embrapa.br/en/web/portal/about-us (Accessed: 11 August 2020).
11 Flor Rafael, F. (2019) "Focusing on 'protective foods' to reduce the global burden of disease," *Rockefeller Foundation Blog*, 24 April [Online]. Available at: https://www.rockefellerfoundation.org/blog/focusing-protective-foods-reduce-global-burden-disease/ (Accessed: 26 August 2020).
12 The Food and Land Use Coalition. (2019). *Growing better: Ten critical transitions to transform food and land use* [Online]. Available at: https://www.foodandlandusecoalition.org/wp-content/uploads/2019/09/FOLU-GrowingBetter-GlobalReport.pdf (Accessed: 25 August 2020).
13 EAT. (2019). *Healthy diets from sustainable food systems: Food planet health* [Online]. Available at: https://eatforum.org/content/uploads/2019/01/EAT-Lancet_Commission_Summary_Report.pdf
14 Consumer Brands Association. (n.d.) *Home page* [Online]. Available at: https://www.gmaonline.org/downloads/research-and-reports/consumerproductfraud.pdf (Accessed: 24 August 2020).
15 The Citizen. (2016). "Over half of imported goods used in Tanzania are fake," *The Citizen*, 17 June [Online]. Available at: https://www.thecitizen.co.tz/news/Over-half-of-imported-goods-used-in-Tanzania-are-fake/1840340-3253536-dkravo/index.html. (Accessed: 24 August 2020).
16 UNICEF. (2020) *Malnutrition* [Online]. Available at: https://data.unicef.org/topic/nutrition/malnutrition/ (Accessed: 24 August 2020).
17 Food and Agriculture Organization of the United Nations & World Health Organization. (n.d.) *Codex Alimentarius* [Online]. Available at: http://www.fao.org/fao-who-codexalimentarius/about-codex/en/#c453333. (Accessed: 24 August 2020).
18 World Spices Congress. (2010) *Food safety and risk assessment by South African regulators* [Online]. Available at: http://worldspicecongress.com/uploads/files/13/WSC10TOPIC08.pdf

    Health24. (2017) *Sudan red: What you should know.* [Online]. Available at: https://www.health24.com/Medical/Digestive-health/Gastroenteritis-and-food-illness/Sudan-Red-what-you-should-know-20120721.

    Food Advisory Consumer Service. (2019) *Sudan red* [Online]. Available at: https://foodfacts.org.za/sudan-red/ (Accessed: 24 August 2020).
19 Ghana News Agency. (2017) "Be wary of food fraud and unsafe items – FDA," *Ghana News Agency*, 29 October [Online]. Available at: https://www.ghanaweb.com/GhanaHomePage/NewsArchive/Be-wary-of-food-fraud-and-unsafe-items-FDA-595106 (Accessed: 24 August 2020).

20 Japan External Trade Organization. (n.d.) *Home page* [Online]. Available at: https://www.jetro.go.jp/en/. (Accessed: 11 August 2020).
21 Nippon. (2018). "Number of overseas Japanese restaurants tops 100,000," *Nippon*, 15 June [Online]. Available at: https://www.nippon.com/en/features/h00218/number-of-overseas-japanese-restaurants-tops-100-000.html (Accessed: 11 August 2020).
22 Kloman, H. (n.d.) "Find an Ethiopian restaurant," *Ethiopian Restaurant Guide* [Online]. Available at: https://ethiopianrestaurants.wordpress.com/ (Accessed: 30 August 2020).

8

BUILDING RESILIENCE – ADAPTING TO CLIMATE CHANGE, MITIGATING RISKS, AND SHOCKS

Introduction

Research indicates that 43% of businesses never reopen after a disaster, and 25% of the companies that do, fail within a year.[1] The lockdowns and social distancing restrictions imposed by governments in response to the COVID-19 pandemic crippled local and international trade and led to substantial declines in purchasing power from job losses and reductions in remittances. Farmers faced challenges accessing inputs, such as seeds and fertilizer, at the beginning of the planting season and reaching markets for their produce. In countries such as Nigeria, Gabon, and Angola, the oil price shocks that coincided with the beginning of the COVID-19 pandemic led to local currency devaluations and a significant contraction of the economy, further worsening the situation. These shocks further exposed the fragility of the African food ecosystem and led to the partial and full closure of many businesses.

The sad reality is that risks, shocks, and crises are the new normal in the food and agriculture landscape. Clearly, we can only expect more human-made and natural disasters linked to climate change, future pandemics, economic shocks, and social crises. This chapter will highlight strategies for building your resilience as a business leader and provide you with systems and structures for strengthening your company's ability to withstand future crises.

Building resilience as a business leader

A 2020 publication by Deloitte[2] described resilient leaders as

> genuinely, sincerely empathetic, walking compassionately in the shoes of employees, customers, and their broader ecosystems. Yet resilient leaders must simultaneously take a hard, rational line to protect financial performance from the invariable softness that accompanies such disruptions. Resilient leaders take decisive action – with courage – based on imperfect information, knowing that expediency is essential.

The lessons from entrepreneurs who continued to thrive despite COVID-19 and climate change shocks reveal that they possessed a few critical attributes and values outlined in this quote. More specifically, they had faith that they would survive the crises and even would thrive through it. They were life-long learners who acquired knowledge and, as a result, could anticipate risks and effectively mitigate or adapt to them. They could clearly articulate their vision and passion for the company before, during, and after the crises. These entrepreneurs were agile, ready to respond to disasters, and prepared to work with others in the ecosystem to solve problems effectively. They demonstrated empathy in terms of how they engaged with their team members, suppliers, distributors, and customers, giving them confidence about their effectiveness as leaders and showing that they cared about these stakeholders in both the short-, medium-, and long-term. In addition, these entrepreneurs were effective communicators and had strong interpersonal and relationship management skills.

> **Box 8.1: SNAPSHOT OF COMPANY RESILIENCE IN APRIL/MAY 2020 DURING COVID-19 FROM THE NOURISHINGAFRICA.COM HUB**
>
> **Farm Fresh (Gambia)**
>
> Farm Fresh is the first physical and online (delivery) fruit and vegetable store in Gambia.
>
> While the company experienced a drastic reduction in in-store customers, their online sales and delivery increased by 400%. Led by founder Modou Njie, Farm Fresh also created a series of popular packages including the "Stay Home Package," which included 14 staple and popular products (i.e., rice, sugar, potatoes, onions, cooking oil, tomato paste, corned beef box, sardines, powdered milk, tea, coffee, mayonnaise, and butter); and the "Immune System Booster Package" comprising of moringa powder and assorted fruits.
>
> **Shambapro Limited (Rwanda)**
>
> Shambapro is an agtech company that provides farmers with access to quality farm inputs. Through its online platform, farmers can search, order, and pay for farm inputs by Mobile money and have their purchases delivered to them anywhere in the country.
>
> During the COVID-19 crisis, once agriculture was deemed an essential service in Rwanda, founder, Kelvin Odoobo, saw a new selling opportunity for the company's e-commerce platform, aggressively turning to popular social media platforms in Rwanda, including Twitter, Facebook, and WhatsApp, to promote its products, direct more traffic to its portal, and engage with potential clients who were now spending most of the time online because of the lockdown. Based on the feedback and validation of this tactic, Shambapro proceeded to create and launch a new product called Shambapro Urban Home Garden Kits – a packaged bundle of seeds and services delivered to urban homes aimed at helping them engage in farming and serve an assortment of fresh organic vegetables on their tables. Shambapro now supplies everything needed for the customers to set up an urban farm in their homes, help them set up, and give them online management tips.
>
> **Agrimod Solutions Centre (Kenya)**
>
> Agrimod Solutions Center is a Kenyan based organization that provides specialized services in organic farming systems and technologies to

> agripreneurs toward promoting food security, environmental sustainability, and biodiversity.
>
> Because of the need for social distancing, Agrimod's expert trainers shifted focus from field training to online conferencing via Microsoft Teams because of the pandemic. The Agrimod experts interact with agripreneurs over some time per day, sustaining its client base and driving new interests. According to Agrimod founder David Adeoye, "the response is so overwhelming, and we are certain to carry on beyond the pandemic." Agrimod Solutions has been able to reach a wider audience because of its move to digitizing operations.

In addition to clear attributes and values, entrepreneurs who thrived during the COVID-19 pandemic took decisive actions in response to the shocks.

- **They ensured the health and well-being of all employees and other critical stakeholders:** This demanded a change in workstyles and workplace engagement, from shifting to remote work and embracing digital tools for communicating with and serving farmers, customers, and other stakeholders, to changing their entire business models and operations to decongest their factories and ensure social distancing.
- **They stabilized their businesses and adapted to the new realities:** This included renegotiating loans, supplier and distributor agreements, and payment terms to preserve cash and to cut unnecessary costs to ensure business continuity and give confidence to customers and employees. They also pivoted their business model to respond to the changing needs of the consumer or customer and address new gaps in the market. During COVID-19, some palm oil producers started manufacturing soap, using their valuable raw materials, given the increased focus on handwashing. Biscuit manufacturers, recognizing the shift away from snacks given the lower purchasing power of consumers and price sensitivity, introduced bread options.
- **They built resilience into their business:** These entrepreneurs invested in technology, innovation, insurance, and currency hedging strategies. They also engaged in scenario planning, considered worst-, base-, and

best-case scenarios, and developed mitigation and adaptation strategies to mitigate risks.
- **They worked with other industry actors to shape their ecosystems:** Recognizing the urgent need for an enabling policy environment that support entrepreneurs during shocks and crisis, the entrepreneurs worked directly or through industry and sector associations to shape policy transformation and implementation. More specifically, they wrote articles and opinion pieces, engaged in webinars and policy dialogues, and joined committees to ensure that agriculture and food were considered essential and that appropriate policies were developed to support the sustainability and growth of the sector. They also supported governments at the local, state, national, and regional levels to reimagine and plan for a different future and actively tracked the implementation of policies to ensure coherence, transparency, and a level playing field.

As an entrepreneur, you must make a concerted effort to build your resilience as a leader and to imbed risk management into your company's DNA. Complete the worksheet below for an honest assessment of where your company stands on some key indicators.

WORKSHEET 8.1: HOW RESILIENT IS YOUR BUSINESS?[3]

Company Data
Have you:
- Maintained updated emergency contact information for employees, vendors, suppliers, customers, and other key contacts?
- Made accessible all important data or files for decision-making if you were unable to access your facility?
- Organized all your critical documents and information so they are easily accessible when needed most?
- Regularly backed-up your data. Is your data stored in a reliable place? Is your data easily accessible remotely?

Employee Engagement
Have you:
- Put in place systems and structures to protect your employees and keep them safe during a health, social, or economic crisis or a natural disaster?
- Assembled a "team" of individuals within the business who know key operations and can provide important perspectives when planning for and responding to disasters?
- Assigned someone to lead business disaster planning efforts for your business? Do you have trained employees to assist (e.g., respond to injuries, evacuate the building) when an emergency occurs?
- Maintained emergency supplies for your businesses to address immediate needs, such as if employees are unable to go home?
- Encouraged employees to be prepared at home? Do you have the basic medical records of all your employees? Do you have an emergency medical coverage for all your employees?

Risk Management
Have you:
- Identified and prioritized which business operations are critical so you know what to recover first, second, etc.?
- Identified the possible hazards (natural and man-made) that could interrupt your business?
- Developed continuity or emergency procedures so you can continue to provide products or services after a disaster? Put a current continuity/emergency/disaster plan in place?
- Maintained procedures to communicate after a disaster with employees, suppliers, vendors, customers, and the public?
- Taken steps to safeguard against potential damage to your equipment, buildings, or facilities?
- Protected produce/farmlands/inventory/storage from theft, loss, or damage?

Ecosystem Engagement
Have you:
- Established partnerships with other businesses, industry associations, government, and/or community organizations that can serve as resources when the next crisis arrives?
- Do you have the phone numbers of the leadership of these entities?
- Are you aware of what support structures, policies, and programs to help companies in need?
- Can you help shape how these interventions are being designed to help companies such as yours?

Environmental
Have you:
- Identified the environmental risks and opportunities associated with your business in relation to your host community?
- Established the appropriate management system, procedure, and policies to mitigate the risks and enhance the opportunities?
- Identified the appropriate management practices to mitigate and/or absorb the climatic, social, and market shocks of your business/farm operations?
- Put in place a health and safety policy and procedure for your suppliers, vendors, customers, and the public? Is it strictly adhered to?
- Identified gaps to ensure effective adherence?

Social
- Do you have a grievance mechanism in place to receive and address complaints from your employees and other external stakeholders?
- Does the system effectively address all complaints fairly?
- How often do you communicate with your value chain stakeholders?

Building financial resilience

A key reason many businesses die following a crisis is because they cannot meet their financial obligations to their suppliers and employees. While many entrepreneurs recognize that "cash is king," they struggle with managing their financial resources during periods of uncertainty. Following the six steps outlined below will enable you to build your company's financial resilience.

Build and protect your savings buffer: Ensure that you have at least 4–6 months of cash to sustain business operations, assuming that there are no new inflows. This requires that you provide incentives for customers to prepay for products or services and ensure faster cash collection cycles. You must also identify customers with large outstanding balances and aggressively follow-up while instituting clear policies around late payment penalties. In addition, leverage technology to foster online orders and payments to enhance the effectiveness and efficiency of future sales.

When funds are generated, they should be invested in low-risk savings tools such as fixed deposits or money market accounts so that they are easy to access with minimal costs for early withdrawals. Some African banks provide double-digit returns for sizable fixed deposits. Entrepreneurs should actively identify such opportunities to generate some investment income from the cash held in their bank accounts.

If you have significant cash available, consider investing in fixed assets, such as land and buildings, for your business or purchasing stocks and bonds for short-term investments, which could ultimately appreciate and generate funds for future use.

In addition, invest in currency hedging strategies, especially if your business operates in an environment with significant risks of devaluation. This often requires saving in foreign currency and ensuring that some customers also pay for your services in more stable currencies.

Invest in insurance: This includes life, health, and property insurance, with coverage for your factory, office premises, raw materials, machinery, and vehicles. Always identify credible insurance providers, compare insurance policies, and negotiate for reduced premiums. Also, form the habit of updating your asset register and maintaining open communication lines with your insurance representative to ensure that you have a good understanding of the requirements for claims, including timing and documentation.

Actively manage your costs: Ultimately, businesses that survive and thrive are those that have efficient and effective operations and can deliver products and services at a low cost without sacrificing value. As a result, every month, review the fixed and variable costs in your business and work with your team to explore opportunities to reduce or eliminate them. Set clear targets for specific expenses as a percentage of income, including administrative overhead, staff salaries, marketing, etc.

Manage your debt and all cash outflow obligations: This includes renegotiating existing loans and future credit facilities with your bankers. During COVID-19, some entrepreneurs were able to renegotiate the terms of their loans with their financial services providers and engage with their landlords to reduce their rent or extend their lease agreement at no additional costs.

You must also manage your supplier relationships, including requesting for supplier credit, extending accounts payable, and ensuring cancellation or service suspension clauses into all future contracts.

Invest in regular scenario planning: Most entrepreneurs create positive scenarios when planning their year. However, you must invest in regular scenario planning, assessing the best-, base-, and worst-case possibilities from a cost and revenue perspective. Based on the outcomes of this planning, prepare your business to strive for the best. Institute systems and structures to ensure that your company can survive the worst-case scenario from a liquidity and operational perspective.

Actively seek out opportunities for support: Through hubs like Nourishingafrica.com, explore what grants, subsidies, and loans are available and the eligibility requirements every month. During the COVID-19 pandemic, local and international funders provided a range of recovery funds and incentive programs for entrepreneurs in the food and agriculture sector that were ready to pivot their operations. While a considerable number of entrepreneurs were well positioned to maximize these opportunities, because they had invested in instituting the systems and structures required to demonstrate investment readiness, others were not.

WORKSHEET 8:2: ASSESSING YOUR FINANCIAL RESILIENCE

- How many months of operations can your savings buffer cover?
- What investment tools have you utilized to manage this savings buffer?
- How are you managing currency risk?
- How often do you prepare, review, and analyze financial statements, including the profit and loss statement, balance sheet, and cash flow statement?
- Do you actively track key indicators, including payable and receivable days and ratios, such as working capital and stock/inventory turnover?
- Do you prepare and review 3-, 6-, and 12-month cash flow forecasts using best-, worst-, and expected-case scenarios? How does this scenario planning affect the decisions that you take daily?
- How often do you update cash flow forecasts to reflect actual events and monitor ongoing cash positions?
- What steps have you taken to increase cash inflows and minimize cash outflows?
- If sales dropped by 20%, a major customer exits, or an important supplier is unable to deliver, how long can your business endure these losses?
- What systems and structures do you have in place to ensure prepayments for products or services or to minimize bad debt?
- Do you utilize credit checks on new customers, and have you instituted limits for new/existing customers?
- How frequently do you assess your customers based on their timely payments?
- How do you leverage technology to maximize the efficiency of engaging with your customers and ensure prompt payment?
- What insurance coverage does your company currently have in place? Health, life, property, inventory, raw material, car, etc.?
- What systems and structures do you leverage to manage costs?

Adapting to climate change

As discussed in the introductory chapter, of the ten countries considered most threatened by climate change globally, nine[4] are in Africa. The recent floods, droughts, and locust infestations that have plagued most of the continent over the past few years reinforce the reality that climate change can completely reverse the advancements in the agriculture and food ecosystem unless urgent action is taken.

As entrepreneurs in the food and agriculture landscape, we must all become climate champions, driving behavior change in our companies, communities, and the broader ecosystem, as well as embracing innovations and technologies that support climate adaptation in Africa.

This will require that you address the inherent risks in your product offerings, supply chains, and distribution channels. For example, if your company uses plastic packaging for your products, you must consider more environmentally friendly options. The governments of Rwanda and Kenya recently banned plastic bags, and companies were compelled to adapt to this new reality. Similarly, you must invest in creating recyclable products and packaging and revamping your production process to conserve energy, recycle water, and minimize pollution and waste.

In terms of securing your crop yields and supply chain, partner with local research institutions and extension agencies to explore the potential to revive traditional farming practices that are sustainable, regenerate degraded soils, and are tailored to local environments. Farmers like Yacouba Sawadogo[5] of Burkina Faso, who received the Right Livelihood Award in 2018 for his use of indigenous and local knowledge to regenerate degraded soils, demonstrate what is possible when entrepreneurs leverage homegrown solutions to solve problems. Yacouba used "zai" – traditional planting pits for soil, water, and biomass retention to transform deserts into almost 40-hectares of vibrant forests.

In addition, leverage the range of improved seeds, weather forecasting tools, and technology tools outlined in Chapter 4 to prepare for weather changes and minimize your risks. Finally, partner with other key stakeholders in your ecosystem to advocate for decisive action from the local, state, national, and regional governments to promote climate-smart agriculture. This should include appropriate policies and initiatives that protect the environment with clear incentives and disincentives for companies that do not comply.

Enhancing your resilience by becoming an inclusive business

Shocks like COVID-19 often compel entrepreneurs to reexamine their business models – given the changes in the ecosystem and the pressure to provide products and services to customers and clients who are dealing with their own crises. An in-depth assessment of companies that appear to have withstood these shocks reveals that there are inclusive businesses – they provide

> access to goods, services, and livelihoods for low-income and vulnerable people at the base of the economic pyramid. By integrating the base of the pyramid into their value chains and focusing on them as customers, inclusive businesses deliver sustainable development impact[6].

There are two critical components of their business models that set them apart and ensure that they survive and thrive during shocks. First, they have invested in local sourcing from smallholder farmers and have built strong local ecosystems composed of small- and medium-size enterprises (SMEs) to support logistics and other aspects of their operations. As a result, despite the increasing levels of protectionism and restrictions in global trade, their businesses have faced minimal production disruptions. For example, Nigerian Breweries Plc. (Heineken) has benefited from its decades of investment in local sourcing of sorghum and cassava starch, instead of dependency on imported malted barley and its relationships with local SMEs. The company works directly with smallholder farmers and provides logistical support, ensuring raw material availability to keep its factories operational.

Second, these businesses invested in developing products and services to meet the needs of customers at the bottom of the pyramid (BOP). They had also built out extensive distribution channels to ensure widespread availability. These prior investments ensured that they could retain their customers and benefit from strong sales growth during the pandemic when many other competing products were considered unaffordable. Given the rampant job losses and declines in remittances that affected the purchasing power of many households, these companies even gained new customers during the crisis. This proved to be the case for AACE Foods, which has experienced increased demand for its single-serve spice products, such as its jellof rice, fried rice, curry, and all-purpose spices sachets, which cater to BOP consumers.

As outlined on the InclusiveBusiness.Net[7], there are eight critical questions that entrepreneurs must answer to assess whether their business models are inclusive:

1. Does your company reach people at the base of the pyramid?
2. What role do people at the base of the pyramid play in your company's value chain?
3. To what extent are people at the base of the pyramid a focus of the business?
4. Does the business model deliver impact to the base of the pyramid?
5. How profitable and bankable is your business model now or in the future?
6. How does the company perform along environmental, social and governance (ESG) criteria?
7. Is the business model replicable to reach scale and systemic impact?
8. Is the business model innovative in reaching the BOP, providing competitive advantage?

Agriculture and food entrepreneurs in Africa have to work collaboratively with other key actors in the public, private, and nonprofit sectors to promote local sourcing. This will not only ensure healthier diets but also improve the lives of farmers. These interventions can be patterned after Brazil's successful Food Acquisition Programme (PAA) and the National School Feeding Programme (PNAE). These programs link the supply of produce from smallholder farmers to the demand of institutional procurement for food-based safety net and school feeding programs.

Summary

You can only expect more human-made and natural disasters linked to climate change, future pandemics, economic shocks, and social crises. However, your businesses' ability to withstand these shocks is hinged on the choices that you make as a leader today. Indeed, your decision to invest in the systems and structures to strengthen your operational and financial resilience and in building inclusive business models hinged on local sourcing, innovative BOP products and services, and distribution strategies that engage and serve low-income populations will enable your company to survive generations.

Notes

1 US Chamber of Commerce Foundation. (2017) *Resilience in a box: Strengthening communities globally* [Online]. Available at: https://www.uschamberfoundation.org/sites/default/files/media-uploads/Resilience%20in%20a%20Box%20Checklist%20for%20Business%20Preparedness.pdf (Accessed: 25 July 2020).
2 Renjen, P. 2020. "The heart of resilient leadership: Responding to COVID-19," *Deloitte Insights*, 16 March [Online]. Available at: https://www2.deloitte.com/us/en/insights/economy/covid-19/heart-of-resilient-leadership-responding-to-covid-19.html (Accessed: 25 July 2020).
3 Adapted from the US Chamber of Commerce Foundation. (2017) *Resilience in a box: Strengthening communities globally* [Online]. Available at: https://www.uschamberfoundation.org/sites/default/files/media-uploads/Resilience%20in%20a%20Box%20Checklist%20for%20Business%20Preparedness.pdf (24 July 2020).
4 Nugent, C. (2019) "The 10 countries most vulnerable to climate change will experience population booms in the coming decades," *Time,* 11 July [Online]. Available at: https://time.com/5621885/climate-change-population-growth (Accessed 24 August 2020).
5 The Right Livelihood Foundation. (2018) *Yacouba Sawadogo* [Online]. Available at: https://www.rightlivelihoodaward.org/laureates/yacouba-sawadogo/ (Accessed: 24 August 2020).
6 Geaneotes, A., & Mignano, K. (2020) "Leveraging inclusive business models to support the base of the pyramid during COVID-19," *EM Compass*, 84, International Finance Corporation [Online]. Available at: https://www.ifc.org/wps/wcm/connect/c03a2420-d82e-4480-8dd4-e5c90e5a210e/EMCompass_Note+84-LeveragingInclusiveBizModels-COVID_FIN.pdf?MOD=AJPERES&CVID=n8PTSeg. (Accessed: 24 August 2020).
7 Sobhani, S. (2019) "What is inclusive business anyway?," *Inclusive Business*, 4 September [Online]. Available at: https://www.inclusivebusiness.net/ib-voices/what-inclusive-business-anyway. (Accessed: 24 August 2020).

CONCLUSION

"If I tell you my dream, you might forget it. If I act on my dream, perhaps you will remember it, but if I involve you, it becomes your dream too." This proverb completely captures my entrepreneurial journey with LEAP Africa, AACE Foods, Sahel Consulting, and Nourishing Africa. I have actively engaged and involved dynamic young Africans, and they have become the greatest champions, advocates, and supporters of the sector.

The essence of this book is to involve you as well in taking full ownership of transforming the food and agriculture landscape in Africa and ensuring that we work collaboratively to achieve our ambitious dreams – as captured in the vision of nourishingafrica.com.

By 2050 – Africa will have a flourishing, sustainable, and just food ecosystem that leverages agtech and digital innovations, driven by Africa's vibrant entrepreneurs to ensure that the continent nourishes itself and becomes a net exporter of food.

I am excited that other entrepreneurs across Africa already share many aspects of this dream. Daniel Ndaka Sila, the Kenyan finalist for the Rockefeller Foundation's Food Systems Prize focused on envisioning regenerative and nourishing food futures for 2050, has outlined a very bold and courageous vision for the city of Nairobi:

By 2050, the urban residents of Nairobi City will be served by a food system characterized by: (a) an efficient, diversified, and sustainable food production system; (b) intelligent distribution interface; (c) smart food processing and preparation options; and (d) personalized food service and diets. By 2050, people will eat food because they like it. While culture, diet, economics, and environment will be important, science, technology, and innovation will be pivotal in driving food production systems, the related delivery systems, and food choices. Artificial Intelligence will form the basic hallmark for food choices and preparations. It is foreseen that Nairobi will require developing and implementing enabling policies and guidelines to accelerate attainment of food and nutrition security for better health. It is in this respect that JKUAT will work with Nairobi County, KEBS, Government ministries (e.g. Ministries of Health, Agriculture, Industrialization, Trade, Environment etc.) industry actors, NGOs, development partners among other stakeholders, to make this vision a reality.[1]

The attainment of the visions for Nourishing Africa and the City of Nairobi are hinged on all the critical issues addressed in this book – the need for compelling and sustainable business models, talent, innovation and technology, brand, financing, and the active participation of ecosystem actors in public, private, and development sectors.

I have two final critical messages:

For policymakers and development partners: The COVID-19 pandemic, the locust infestation in East Africa, and droughts in many countries on the continent have further highlighted the fragility of the African food ecosystem and the urgent need to look inward to invest in ensuring food self-sufficiency. These shocks have also reinforced the importance of human diets in boosting immune systems and ensuring food security and nutrition.

The reality is that there will be future shocks linked to climate change. As critical stakeholders, we must collaboratively and urgently rebuild the African agriculture and food landscape. This will require a concerted effort to support Africa's youth entrepreneurs engaged in the agriculture landscape. We must invest in private-public partnerships to create an enabling environment for enhanced value chain competitiveness, transitioning our subsistence farmers to thriving agribusiness entrepreneurs, leveraging digital tools and climate adaptation strategies. We must also unlock catalytic financing mechanisms and infrastructure, including roads, rail, storage, and sustainable energy solutions, which are critical to reducing postharvest losses and fostering local value addition and a vibrant food ecosystem. In addition, we must invest in food safety and high standards for our products and build more robust market and trade linkages on the continent, leveraging the African Continental Free Trade Agreement (ACFTA) to ensure greater regional and international trade. Finally, we must prioritize the agency, resources, assets, and skills for women as critical stakeholders in this ecosystem.

For African entrepreneurs in the food and agriculture landscape: Our time is now! We must take our rightful place in the entrepreneurial ecosystem – be bold, brave, and resilient, leveraging innovation and technology to propel our sector as Africa's engine of growth. With mindsets rooted in humility, a commitment to life-long learning, collaboration and service, and the requisite skills and business models, I am confident that we can create this $1 trillion industry and, in our lifetime, achieve the dream of Africans nourishing our continent and the world.

Note

1 Ndaka Sila, D. (2020) "Nairobi Food System Vision 2050 for improved nutrition and better health in Kenya," *Food System Vision Prize*, 20 June [Online]. 2020. Available at: https://challenges.openideo.com/challenge/food-system-vision-prize/refinement/nairobi-food-system-vision-2050-for-improved-nutrition-and-better-health-in-kenya (Accessed: 26 August 2020).

INDEX

Note: **Bold** page numbers refer to tables and *italic* page numbers refer to figures.

AACE Food Processing & Distribution Ltd. xxi, 12–14, 18–19, 34, 35, 74, 85–86, 117, 128, 144–145, 152, 171–172, 174
accelerators 113–115
Access Agriculture Young Entrepreneur Challenge Fund 117
Acholonu, U.K. 35
Acumen 129
Acumen East Africa Fellows Program 44
Adegoroye, Temi 95
Adeoye, David 163
"adjacency strategy" 16–17
AFEX 72
affected/influential bystanders 136
Africa Agriculture Status Report (2019) xxi
Africa Enterprise Challenge Fund (AECF) 117
Africa Improved Foods (AIF) 20, 146–147
African Angel Business Network (ABAN) 112
African Continental Free Trade Agreement (ACFTA) 176
African Enterprise Challenge Fund (AECF) 127
African Regional Economic Communities 153
African Union 153

aggregation 71–74
agility 30
AgriBio 148
Agribiz4Africa 115
agribusiness 18, 29, 57, 154; scaling in 55; youth engagement in 8
Agribusiness Africa Window (AAW) 128
agribusiness education transformation 154
agricultural ecosystems xxi
agricultural inputs 62; crop protection 66–67; delivery of 67; mechanization 67–68; seed technologies 62–65; soil health and fertility 65–66
Agriculture Africa Accelerator 114
Agriculture and Climate Risk Enterprise Ltd. (ACRE Africa) 71
agriculture and food business: ecosystem changes required in 147–149, 150, 151–156; founders and management teams of 46; roles in 38–39; scaling in 37
AgriFI Kenya Challenge Fund 117
Agrifood SMEs, reasons VC/PE funds decline applications from 123–124
agrifood tech companies 75
Agrikore 71
Agrimod Solutions Centre (Kenya) 162–163

INDEX

agripreneurs 43
AgroSpaces 69
Agro Supply 114
agtech 60–61, **61**
agtech business service provider: attracting, retaining, and scaling paying customers/subscribers 76–78; compelling product/service offering and business case 75–76; measuring impact/value creation 78–79
AIDAR model 100, 104
Ajene, Nwando 35
Ajogwu, Fabian 35
Akellobanker 114
Akinboro, Bolaji 71
Akumah, Onyeka 71
Alliance for a Green Revolution in Africa (AGRA) xxi
alternative schedule 55
angels 112
Arifu 70
Artificial Intelligence 175
Australian Export Grains Innovation Centre (AEGIC) 147–148
automated technologies 68
award(s) 115, **116**, 128
Aywajieune 114

Babban Gona 69, 112, 120
Baby Grubz 87
Baird, Ross 37
Bakoume, Jim 69
balance sheets 124
Balogun, Ayodeji 72
Beat Drone 69
behavioral segmentation 90
beneficiaries 136
benefits 49; of embracing innovation and technology 58–61, 59
benefit segmentation 90
BIF *see* Business Innovation Facility
big data, crop production 70–71
Bill & Melinda Gates Foundation 112
Biofortification 64–65
Black Mamba Foods 37
blockchain technology 68
Bloom, Paul N. 136
blueMoon 114
board members of nonprofits 33–34
board of directors, establishment of 33–35
bonuses 49–51
Bosire, Peris 138

bottom of the pyramid (BOP) 171–173
brand 83; company name 84–86; designing logo and colors selection 86–87, 87; financing 106–107; promise 88–90; strategy steps for developing 84–90; tagline 87
brand building approaches 105
branded gifts 95–96
branding positioning 93
Brazilian Agricultural Research Corporation (Embrapa) 148–149
Brooke, Grant 13, 18
Brookside Dairy Ltd. 98, 99
Brown, Ross 135
Bruni, Michele 72
budgets 124
Building an Economically Sustainable Integrated Seed System for Cassava (BASICS) 63–64, 96
bundling 96–97
Burkina Faso 170
Business Angels 112
Business Innovation Facility (BIF) 117
business leader, resilience as 161–166
business models 171, 172; demand-driven with measurable value addition 24–25; measurable impact 26–27; prerequisites for 24; scaling 23–27
business-to-business 18–19; classification 19; sales 19
business to consumer 19–20
business-to-government sales 20

Cameroon Angel Network (CAN) 112
Candel Company 66–67
capital, issues assessed by providers of **122–123**
Capture 114
cash flow statements 124
cash outflow obligation 168
Castellanos, Claudia 37
CCHUB 114
Cellulant 71, 119
CGIAR 70
challenge funds 117
Chamber, Tim 72
City of Nairobi 174, 175
climate change 2, 170
CNRA *see* National Centre for Agricultural Research
cocoa, in Ghana 6
CODEX 153

cofounders 36–37
ColdHubs 72
commercial banks 120
commodity exports 29
communications 49
company culture 54–55
company name, selection of 84–86
compensation 49
compensation planning 48
Competition Act 139
competitions 115, **116**
competitive landscape 139–140
competitors 136
Complete Farmer 114
consultants, short-term 45–46
Consumer Goods Council of South Africa (CGCSA) 142
Context Development Network 96
cost management 60–61, 168
Cote D'Ivoire 148
country of interest 16–17
COVID-19 pandemic xxii, 32, 57, 58, 138, 160, 161, 175; actions in response to shocks 163–164; company resilience in April/May 2020 162–163; resilience by becoming inclusive business 171–172
Cowtribe 70
Creadev 119
Critical Capital for African Agrifood SMEs report 123
critical realities 1–5, 4
crop production: big data 70–71; extension support/service provision for farmers 69–70; financing farmers using digital tools 71; range of innovations available for 68–69
crop protection 66–67
cross-border trade 117
crowdfunding 115

Dalberg 58
Danso, Evans 15, 51
debt management 168
decline, financing 121, **122**–123, 123–131, **128**
Dees, J. Gregory 136
Deloitte 161
Demand Creation Trials (DCTs) 96
demand-driven, with measurable value addition 24–25
demand-driven research 147–149
demographic segmentation 90

development partners 175–176
diets 8
digital branding 108
digital innovations, momentum of 7
digital interventions 59
Digitalisation of African Agriculture Report, 2018-2019 7
digitalization for agriculture 58
digital marketing 103, 105
digital media trifecta model 103
digital skills 31–33
digital tools 60–61, **61**; financing farmers using 71; through value chain 61
displays 97–98
Dlamini, Nhlanhla 13, 15
double standards, inequity in 5
Draper Richard Kaplan Foundation 129

Early Generation Seed (EGS) System 25, 62–64
earned media channels 103
East Africa, cross-border trade 117
EatFresh 94
Ecodudu, Kenya 65
e-commerce food store 71
ecosystem 134, 164; changes required in African agriculture and food landscape 147–149, 150, 151–156; competitive landscape 139–140; culture and norms of society 144–145; forming strategic and sustainable partnerships 145–147; geography and infrastructure 140–142; mapping 136–137; professional networks and associations 142–144; regulatory environment 137–139
eeZee Noodles 98–99
Embrapa 148–149
emotional intelligence 30
employee relations 48
employment 8
eMsika 114
enablers/drivers 136
entrepreneurial ecosystem 135, 135–136
entry-level employees 41
Enviro-Gro Farms Limited 36
equity 50, 125; focus on 8–9
equity-based crowdfunding 115
equity financial landscape 29
Ethiopian Commodity Exchange (ECX) 72–73
Ethiopian cuisine 156
export-focused value chains 15

FAFIN *see* Fund for Agricultural Finance in Nigeria
Fairfield Dairy 130
Famae Food Challenge 117
Farmcrowdy Ltd. 71
FarmDrive 138, 140
farmer field days 96
farmers: extension support/service provision for 69–70; financing using digital tools 71
Farm Fresh 74–75, 162
Farmshine 119
fast-moving consumer goods (FMCGs) 97
FATE Foundation xx, 140
fellows 44–45; short-term consultants 45–46
Fellowship Programs 44
fellowships 113
"Fidelity SME Forum" 95
financial planning and management 31
financial resilience 167–169
financial snapshots 124
financing: decline 121, **122**–123, 123–131, **128**; in food and agriculture landscape 110; in growth phase 118–120; maturity 121; options 111, **111**; in startup and early phase 111–117
financing gaps 3–4
flexible schedule 55
Food Acquisition Programme (PAA) 172
food and agriculture landscape 13, 17, 18, 23, 27; for African entrepreneurs in 176; entrepreneurs in 170; failed partnerships in 36; financing in 110; scaling in 55
Food and Land Use Coalition (FOLU) 149, 150, 151
food ecosystem xxii
food fraud 151–153
Food Fraud Network 153
food processing training programs 29
food system 174–175
food waste 2
Ford Foundation xx
formal contracts 45
"freemium" version 76
free trainings 78
free trials 76–78
Fund for Agricultural Finance in Nigeria (FAFIN) 119–120
funding: plan for exits 130–131; steps for raising 121, **122**–123, 123–131, **129**
fundraising process 125–126

Gabre-Madhin, Eleni 114
Gallup 55
Gambia, Farm Fresh 74–75, 162
gender gap 5
The Gender Gap in Agricultural Productivity in Sub-Saharan Africa: Causes, Costs and Solutions 4
gender inequity 4–5
genetically modified organisms (GMOs) 65
geographic segmentation 90
geography, ecosystem 140–142
Get on Board: A Practical Guide to Building High Impact Board of Directors in Nigeria (LEAP Africa) 33
Ghana: cocoa in 6; palm oil 152
Ghana Angel Network (GAIN) 112
Ghana Food and Drug Authority 153
GIZ Innovation Challenge 117
Global Alliance for Improved Nutrition (GAIN) 72
Global Entrepreneurship Summit 93
global food industry 1
Global Innovation Fund 120
globalization 5
global trade dynamics 5
global trade, inequity in 5
Glow Healthy Smoothies and Snacks 88, 100
Goldman Sachs 119
Good Nature Agro 36, 50, 52, 52
Google AdWords 103
grants 112–113, 126
Gray Matters Capital 119
Green Home Packaging 74
Gro Intelligence 70
growth phase, financing in 118–120

Haile, Martha 84
HarvestPlus 64–65
Hays, Kellan 36
health maintenance organizations (HMOs) 50
healthy diets 149, 150, 151
Hello Tractor 67–68, 84, 93, 94, 103
high-roofed greenhouses 68
human resource management 31
human resources 47–48

ideal boards 33
IItA GoSeed 46–47
incentives 49–51
InclusiveBusiness.Net 172
incubators 113–115
Indomie Noodles 106

inequity, in global trade and double standards 5
infrastructure 3–4; ecosystem 140–142
Initial Concept Note Assessment 127
in-kind support 117
innovation: benefits of embracing 58–61, 59; through value chain 61
Innovations Against Poverty 117
InspiraFarms 72
Institute for Health Monitoring and Metrics (IHME) 149
in-store tasting 97–98
insurance, invest in 167–168
integrity 31
intellectual property (IP) 76
International Finance Corporation 118–119
International Institute of Tropical Agriculture (IITA) 96
Interprofessional Fund for Agricultural Research and Advice (FIRCA) 148
interregional trade 5
investment 129; criteria for 29; in insurance 167–168; in regular scenario planning 168
investors 118–120, 129

Jabiry, Hadija 94
Japanese cuisine 155–156
Japanese External Trade Organization (JETRO) 155
Java Foods Limited 98
Jensen, Carl 36
JKUAT 175

Kenya 7, 9, 170; Arifu 70; Ecodudu 65; export-focused chains 15; Global Entrepreneurship Summit 93; Tropical Heat 91; vegetable oil 152
Kenya Bureau of Standards 153
Kewalram Chanrai Group 16
key performance indicators (KPIs) 49
Kiernan, Shane 115
Kobo360 73, 118–119
Kutukwa, Kudzai 71

labor law compliance 48
Lagos Angel Network (LAN) 112
Lang, Kevin 130
leader, building your capacity as 30–33
LEAP Africa xx, 33, 174
learning culture 54
leave 50

lending-based crowdfunding 115
leveraging precision agriculture 68
life and health insurance 50
life-long learning 50–51
loans 120
locals/expatriates 45–46
logistics 71–74
Lori Systems 72
Lovell Industrial 74
low-performing staff 52–53

Maano app 75
malnutrition xxii, 149, 150, 151; high rates of 2–3; in Nigeria 12
Maneli Foods 13, 15
Maneli Pets 19, 89
market-based competition 139
marketing strategy 90–93; financing 106–107; measuring effectiveness of 108
market linkages 74–75
market storms 97
Masha, Kola 69
Mason, Colin 135
Mastercard Foundation 112–113
McKinsey 53
mechanization, agricultural inputs 67–68
Mehta, Mira 84, 115
middle class 8
millennials, working with 55
mission statement 22–23
Mobbinsurance 71
Mossavar-Rahmani Center for Business & Government xxi
motivation, by management 51
M-Pesa 140
Musonda 98

National Agency for Food and Drug Administration and Control (NAFDAC) 106
National Centre for Agricultural Research (CNRA) 148
National Competition Commission 140
National School Feeding Programme (PNAE) 172
National Union of Coffee Agribusinesses and Farm Enterprises (NUCAFE) 44
Ndaka Sila, Daniel 174
networking skills 31
networks 94
Nigeria 14; AFEX 72; CCHUB in 114; Da-Algreen and Value Seeds in 25;

export-focused value chains 15; maize value chain in 16; malnutrition in 12; Olam Nigeria Plc 16–17; SoFresh 41, 42, 43
Nigerian Breweries Plc. 171
N'jie, Modou 75, 162
Njonjo, Peter 13, 18, 84
Njoroge, Ken 71
Nkandu, Joseph 44
noncommunicable diseases 3
Nourishing Africa 142–143, 174, 175
Nourishingafrica.com xxi, 138, 142–143, 162–163
NUCAFE *see* National Union of Coffee Agribusinesses and Farm Enterprises
Nutzy Peanut Butter 96
Nwuneli, Mezuo 12, 34

Obama, Barack 93
obesity 3
OCP 66
Odoobo, Kelvin 162
Okpareke, Nkiru 36
Olam Nigeria Plc 16–17
Oliver, Jehiel 93, 94
onboarding 42–43
One Acre Fund 26, **128**, 129
online presence, traditional marketing approaches 100, 101–104, **102–103**, 104
operating costs 59
opponents/critics 136
opportunity 14, 15; for support 168
organizational charts/structure 43
organizational culture, high-performance and conductive 47–49
outsourcing 46–47, 126
overweight 3
owned media channels 103

packaging 91–92
packaging solutions 74
paid media channels 103
Pareto principle 20
Partners in Food Solutions 44
partners/partnerships 36–37, 96–97; strategic and sustainable 145–147
part-time schedule 55
PEAs *see* Private Extension Agents
peer-based feedback 51
people 92
performance evaluations 51–53, 53

personal discipline 31
Phatisa 130
physical marketing 105
place 92
policymakers 175–176
Postharvest Loss Alliance for Nutrition (PLAN) 72
postharvest losses 2
postharvest storage/aggregation/logistics 71–73; processing 73–74
price 91
primary region 16–17
print media 95
private equity (PE) 118
Private Extension Agents (PEAs) 50, 52
prizes 115, **116**
processing, value chain 73–74
product 91
professional networks and associations 142–144
profit and loss statements 124
promising trends 7–9
promotion 49–51, 92, 97–98
psychographic segmentation 90
public-private partnership 20
public relations (PR) 105

QuickTrials 76–77

racism 8
radio 94–95
Raino Tech4Impact 114
ransparent culture 54
recognition, by management 51
recruitment 40, 48
Reelfruit 114, 126–127
referrals 40
regional trade dynamics 5
regulatory environment 137–139
repayable grants 120
repositioning African food globally 154–156
Request for Proposal (RFP) process 45
resilience 30; by becoming inclusive business 171–172
resilient leaders 161–166
restricted grants 112
retail 75
reward-based crowdfunding 115
Right Livelihood Award (2018) 170
risk management 60–61

roadshows 97
robust recruitment process 40
Rockefeller Foundation's Food Systems Prize 174
Roques, Joe 37
route to market (RTM) 98–99
Rwanda 170
Rwanda-based Africa Improved Foods 97

Sahel Capital Agribusiness Managers Limited (SCAML) 119–120, 129
Sahel Consulting Agriculture & Nutrition Ltd. xxi, 40, 49, 51, 54, 95, 96, 152, 174; maize value chain in Nigeria 16; Quarterly research 3, 4
Saillog 69
Sandler, Joshua 72
savings, build and protect 167
Sawadogo, Yacouba 170
"Scale Innovations" 26
Scaling Up Nutrition (SUN) Movement 143
SCGN *see* Society for Corporate Governance Nigeria
seed technologies 62–65
sensors 68
service delivery 59
Shambapro Limited (Rwanda) 162
SheLeadsAfrica 115
short-term consultants 45–46
SIDA *see* Swedish International Development Cooperation Agency
Silungwe, Sunday 36
Singapore Stock Exchange 17
six Ps 90–92, 92, 108
skill-will matrix 53
small- and medium-size enterprises (SMEs) 44, 103, 106, 118, 138, 171
smallholder farmers 2, 171
Social Innovation in Africa: A Practical Guide for Scaling Impact 23, 136
social media 8, 108; strategy 104
society, culture and norms of 144–145
Society for Corporate Governance Nigeria (SCGN) 35
SoFresh 41, 42, 43
soil health and fertility 65–66
staff salaries 49–51
startup and early phase, financing in 111; accelerators and incubators 113–115; angels 112; challenge funds 117;
competitions, prizes, and awards 115, **116**; crowdfunding 115; fellowships 113; grants 112–113; in-kind support 117
starvation xx
State of the World's Children 2019 report (UNICEF) 2
"Stay Home Package" 162
strategic partnerships 145–147
strategic planning 31
Sudan IV 152
SUN Business Network 143
Sustainable Development Goals (SDGs) 130–131
sustainable partnerships 145–147
Swedish International Development Cooperation Agency (SIDA) 117
sweet spot, finding your 13–21, 16

tagline 87
talent 3–4, 37–38; attracting talent 40–41, 42, 43; onboarding 42–43; organizational charts/structure 43; recruitment 40
talent management 48, 52
target customer 18, 20–21; business to business 18–19; business to consumer 19–20; business to government 20
taxes 124
Technical Centre for Agricultural and Rural Cooperation (CTA) 58
technical skills 38
technological advances, yield gaps 7
technology: benefits of embracing 58–61, 59; skills 31–33; through value chain 61
telecommuting 55
ten critical transitions 149, 150
Thiam, Pierre 156
13th month salary and performance bonuses 50
TikTok 104
"TIME 100 List" 156
TLcom Capital 119
Tolaram Group 105
Tomato Jos 84–85, 115
Tony-Uzoebo, Emeraba 36
TradeMark East Africa Challenge Fund 117
"Trade Transparency Solutions" 74
traditional marketing approaches 93; branded gifts 95–96; champions 94; farmer field days 96; in-store tasting,

promotions, and displays 97–98; networks 94; online presence 100, 101–104, **102**–103, 104; partnerships/bundling 96–97; print media 95; radio 94–95; roadshows/market storms 97; route to market 98–99; spokesperson 93–94
training 48, 50–51
Tropical Heat 91
TruTrade Africa 74
Twiga Foods 13, 18, 84

"Uber for Tractors" 67
UK Department for International Development (DFID) 117
Umeron, Patricia 35
UNESCO Intangible Cultural Heritage List 155
United States' Grocery Manufacturers Association 151
unrestricted grants 112
unusual products 105–106
USAID Africa Trade and Investment Hub 117

value chain 2, 7; of interest 14–15; processing 73–74; role within 15–16; technology, innovation, and digital tools through 61
values 22–23
VentureBurn 130
Venture Capital for Africa (VC4Africa) 112

vertical farming 68
Vijayvargiya, Guarav 137
Viktoria Ventures analysis 9
Village Capital 114
Virtual Farmers Market (VFM) 75
visibility 58–59
vision statement 22–23
volunteers 43–44

Wandrag, Stiaan 139
WeFarm 69
West African research institutions 148
Willett, Walter 149
Williams, Affong 126
WIN Industries 90
work culture 54
workplace safety 48
World Bank 8
World Bank Report 139
World Food Prize Foundation 115
World Food Programme 20

Yolélé Foods 156
Youn, Andrew 128
youth engagement, in agriculture 8

Zambia 15, 20; Good Nature Agro 52; Java Foods in 45, 98; Maano app 75
Zambian for-profit social enterprise 36
Zebu Investments 130
Zenvus 68–69